Integrated and Collaborative Product Development Environment

Technologies and Implementations

SERIES ON MANUFACTURING SYSTEMS AND TECHNOLOGY

Editors-in-Chief: Andrew Y. C. Nee *(National University of Singapore, Singapore)*
J.-H. Chun *(Massachusetts Institute of Technology, USA)*

Assistant Editor: S. K. Ong *(National University of Singapore, Singapore)*

Published

Integrated and Collaborative Product Development Environment

Technologies and Implementations

W D Li
University of Bath, UK

S K Ong
National University of Singapore

A Y C Nee
National University of Singapore

W **World Scientific**

NEW JERSEY · LONDON · SINGAPORE · BEIJING · SHANGHAI · HONG KONG · TAIPEI · CHENNAI

Published by

World Scientific Publishing Co. Pte. Ltd.

5 Toh Tuck Link, Singapore 596224

USA office: 27 Warren Street, Suite 401-402, Hackensack, NJ 07601

UK office: 57 Shelton Street, Covent Garden, London WC2H 9HE

British Library Cataloguing-in-Publication Data
A catalogue record for this book is available from the British Library.

Series on Manufacturing Systems and Technology — Vol. 2
INTEGRATED AND COLLABORATIVE PRODUCT DEVELOPMENT ENVIRONMENT
Technologies and Implementations

ISBN 981-256-680-5

Printed in Singapore by B & JO Enterprise

To our families

Our families endured the long hours we spent on this book.
Without their support, this book would not be a reality.

Preface

With the increasing globalisation trend and keener competition in the market, modern product development companies are experiencing much higher pressure on minimising both the design cost and the lead-time. With the rapid advances in computing and Internet technologies, an Integrated and Collaborative Environment (ICE), which is based on the complementary functions of concurrent engineering and Internet-based collaborative engineering, is imperative for an establishment to facilitate the product realisation processes. Enabling technologies such as the feature technology for Computer-Aided Design (CAD), the optimisation technology for Computer-Aided Process Planning (CAPP) and collaborative design technology are being actively investigated.

Feature and CAPP technologies have been studied extensively and major monographs in these areas were published in the early and mid 1990s. Most of the operations in commercial 3D CAD systems are feature-oriented. For the more challenging issues, such as integration and conversion of features for different applications, research works only appear in the last ten years, and the relevant system applications are far from satisfaction. In Internet-based collaborative design, the relevant research and development are evolving dramatically and the trend is driven by the rapidly emerging intelligent and information technologies. However, developed methodologies and systems are still in their infancy.

From early 1999, the authors initiated several projects to investigate the relevant topics, and some research results are presented in this book. The authors summarise the state-of-the-art technologies and present some development paradigms, aiming to provide an overall picture and several examples to the readers. The major contents cover two aspects: (1) The most recent research and development of the aforementioned technologies in the last decade are surveyed, and (2) Details of the implementation strategies and case studies of some prototype systems are illustrated to help the readers understand the underlying algorithms and infrastructures of developing an ICE.

The organisation of the book is as follows. Chapter 1 is a general introduction of the main objectives and themes of this book. This book is organised into two major threads. Chapters 2 − 5 cover three major technologies for establishing an integrated product development environment, namely, the manufacturing feature recognition technology, feature conversion technology based on design-by-feature systems, and the CAPP optimisation technology. In Chapters 6 − 8, the product development environment is extended to support collaborative design activities with the incorporation of the state-of-the-art Internet technologies. Two major technologies are discussed, i.e., collaborative feature-based design technology and Web-based CAPP optimisation technology. For these technologies, the current research and development status are comprehensively surveyed in Chapters 2 and 6 respectively, and the detailed implementations are explained in Chapters 3, 4, 5, 7 and 8.

This book can be used as a text or reference for mechanical/ manufacturing/computer engineering graduate students and as a reference for researchers in the field of concurrent engineering, collaborative engineering and intelligent manufacturing, as well as practising engineers in charge of the utilisation, deployment and development of concurrent and collaborative software tools to support product design and development, etc.

During the development of the projects, the authors have received invaluable support from many colleagues from the Singapore Institute of Manufacturing Technology (SIMTech), National University of Singapore (NUS), Data Storage Institute (DSI) and the Institute of High Performance Computing (iHPC). The authors would specially like to acknowledge the contributions from Associate Professor J.Y.H. Fuh and Dr Z.M. Qiu from NUS, and Dr Z.J. Liu from DSI in the form of discussion, advice and encouragement. The authors are grateful to the editors of the World Scientific Publisher for their patience, encouragement and professionalism during the editing process.

- W.D. Li
- S.K. Ong
- A.Y.C. Nee

Contents

Abbreviation

AAG	Attributed Adjacency Graph
AI	Artificial Intelligence
ANN	Artificial Neural Network
API	Application Programming Interface
IDL	Interface Definition Language
ART	Adaptive Resonance Theory
ASPs	Application Service Providers
BIFS	Binary Format For Scenes
BP	Back-Propagation
B-Rep	Boundary Representation
CAD	Computer-Aided Design
CAE	Computer-Aided Engineering
CAM	Computer-Aided Manufacturing
CAPP	Computer-Aided Process Planning
CE	Concurrent Engineering
CGI	Common Gateway Interface
CORBA	Common Object Request Broker Architecture
CSG	Constructive Solid Geometry
DFM/A	Design for Manufacturability and Assemblability
DSG	Destructive Solid Geometry
EAAG	Enhanced Attributed Adjacency Graph
ERP	Enterprise Resource Planning
FIPER	Federated Intelligent Product EnviRonment
FLGs	F-Loop Graphs
F-Loops	Face Loops
GA	Genetic Algorithm
GEFG	Generalised Edge-Face Graph
GT	Group Technology

HP	Hamiltonian Path
HTML	HyperText Markup Language
HTTP	HyperText Transfer Protocol
ICE	Integrated and Collaborative Environment
ISO	International Standards Organisation
IT	Information Technologies
J2EE	Java 2 platform Enterprise Edition
JDBC	Java DataBase Connectivity
JMS	Java Message Service
JCA	Java Connector Architecture
JNI	Java Native Interface
JSP	JavaServer Pages
KQML	Knowledge Query and Manipulation
LMM	Latin Multiplication Method
LOD	Level-of-Detail
MAM	Master Assembly Model
MAS	Multi-Agent System
MLFF	Multiple-Layer Feed-Forward
OMG	Object Management Group
PDM	Product Data Management
PLM	Product Lifecycle Management
RMI	Remote Method Invocation
RMI-IIOP	RMI over Inter-ORB Protocol
SA	Simulated Annealing
SAM	Slave Assembly Model
SE	Sequential Engineering
SOAP	Simple Object Access Protocol
STEP	STandard for the Exchange of Product model data
TAD	Tool Approach Direction
TS	Tabu Search
VRML	Virtual Reality Mark-up Language
W3D	Web 3D
X3D	eXtensible 3D
XML	eXtensible Markup Language

Chapter 1

Introduction

Concurrent engineering and collaborative engineering are modern philosophies to reengineer the traditional product design and development processes leading to better quality, shorter lead-time, more competitive cost and higher customer satisfaction. A full-scale implementation of the philosophies to form an integrated and collaborative product development environment includes two aspects: (1) the upstream design and downstream manufacturing processes are seamlessly linked through the implementation of some intelligent reasoning and integration strategies for effective information exchange, and (2) collaborative design, fuelled by cooperation strategies and the Internet technologies, is realised through collocating a multi-disciplinary design team and emphasising interpersonal aspects of the group work.

In this chapter, the motivations, concepts and research issues for the establishment of an integrated and collaborative environment for product development are introduced. The enabling Artificial Intelligence (AI) and Internet technologies for system implementation are discussed.

1.1 Concurrent and Collaborative Engineering

In traditional design and manufacturing companies, the product development process is usually organised in a sequential manner, i.e., Sequential Engineering (SE). In SE, specialists from different domains that are involved in product design and the related manufacturing processes work in an isolated and independent way. A stage in product development can be basically thought of as being a black box, which inputs are the results of the activities from the previous stage, and which outputs are taken as inputs for the next stage, as shown in Fig. 1.1. The shortcomings of SE include high developmental costs, expensive re-designs and re-work, poor communications and interactions between designers/process planners/manufacturing engineers, the lack of critical decision-making with respect to all of the product development, etc. [Ranky, 1994].

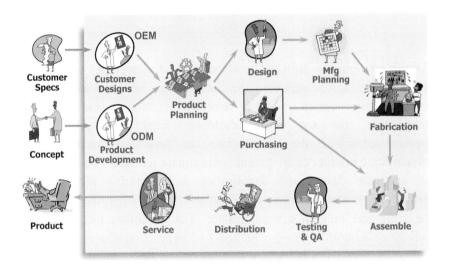

Fig. 1.1 The sequential development process of products.

In a product development process, design concepts and models usually require dynamic adjustments to achieve better product quality and processes. Changes made during the early design stage do not cause significant increase in cost, while during the production stage, sharp increase in cost will incur since many blueprints, design documents or components have been created and these would require re-work and re-design. Subsequently, the lead-time and cost of the product will dramatically increase. This situation is illustrated in Fig. 1.2, which reflects the cost trends of products with respect to the changes in their life-cycle development [Pallot and Sandoval, 1998]. It shows clearly that greater cost reduction opportunities are usually available during the early stages of the product development and therefore it is of paramount importance to identify design and manufacturing problems as early as possible to reduce the risk and cost.

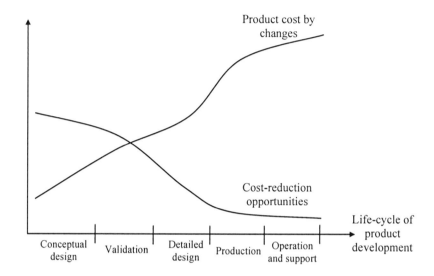

Fig. 1.2 Product cost and cost-reduction opportunities from the life-cycle viewpoint.

Modern product development practices are becoming more product-oriented, and aim at decreasing the lead-time, minimising work-in-process, just-in-time flow of materials, and high efficiency and flexibility of manufacturing capacity utilisation. With the advances in computers and computing technologies, a large number of software tools and philosophies have come into existence to facilitate the product development and realisation processes. From 1990s, the Concurrent Engineering (CE) concept has been adopted by design and manufacturing companies to systematically integrate the design of products with the related manufacturing processes using some software packages and computing technologies in an integrated computing environment. In a CE environment, techniques, algorithms and software tools are provided to allow designers and developers to interact with each other effectively and efficiently. With the implementation of CE, which usually causes more time and money to be spent in the initial design stage to ensure that the concept selection is optimised, companies can reduce design changes at the later stages, leading to better engineered products with an advantage in total quality, time and cost competitiveness (shown in Fig. 1.3).

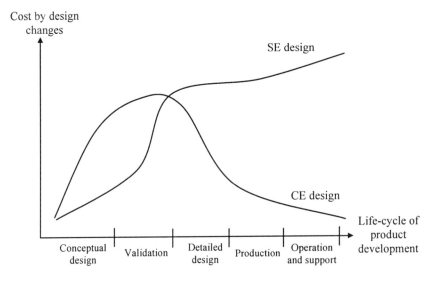

Fig. 1.3 Design changes and the cost incurred in industries.

The CE philosophy can be implemented through different strategies considering the diverse requirements of the users and the conditions of the companies. Pallot and Sandoval [1998] emphasised the communication and sharing of design data, integration and interoperability of the application systems in design and manufacturing, and the coordination between the upstream and downstream activities to support concurrent activities for multi-disciplinary teams. They also investigated some CE related projects that have been funded by the European information technology research program, ESPRIT. The ATLAS project allows the sharing of product models across functional departments based on an international data exchange standard, STEP (STandard for the Exchange of Product model data). Its initiative is to establish a generic data information sharing model for manufacturing enterprises to realise the seamless integration and interoperability of design and manufacturing information. The FIRES project and the SCOPES project focused on the Design for Manufacturability and Assemblability (DFM/A) methodologies respectively to bring manufacturing engineers into the early design decision-making process to optimise the manufacturing and assembling processes when a product is being designed, thereby improving the overall performance of the product. The VEGA project and the MATES project aimed to establish distributed information infrastructures to support remote communication in line with the rapid industrial requirements and the emerging Internet technologies.

From an industrial viewpoint, Balamuralikrishna, *et al.* [2000] emphasised three T's for implementing CE: tools, training and time. Tools refer to the communication facilities between the personals in the multi-disciplinary departments to address the information exchange that is obstructed by the complexity and wide range of specialised disciplinary areas and interdependent activities. One of the greatest challenges in managing the simultaneous operation of inter-related tasks is getting the people to work together as a team. Training provides a mechanism for employees to work collaboratively and concurrently, making the best use of the company resources. Time means corporations need time to carefully investigate and plan CE as it involves many complex software tools and information infrastructures. Many reported cases have shown that a hurried implementation of CE usually brings high probability of backfiring. In summary, a successful implementation of CE needs to consider the following aspects (but not limited to):

- A communication strategy for a multi-disciplinary group of people from the design and manufacturing departments to share and exchange ideas and comments.
- An integration strategy to link heterogeneous software tools in product design, analysis, simulation and manufacturing optimisation to realise obstacle-free engineering information exchange and sharing.
- An interoperability strategy to manipulate downstream manufacturing applications as services to enable designers to evaluate manufacturability or assemblability as early as possible.

With the trend towards global competition and the rapid advances of the Internet technologies, nowadays, extensive research and development have been made towards supporting distributed applications to form a wider landscape, in which geographically dispersed users, systems, resources and services can be synthesised across enterprises in an Internet/Intranet environment beyond the traditional boundaries of physical and time zones. Face-to-face communication and cooperation is impossible in this situation. Some traditional communication methods that are used in a traditional CE environment, such as emails, discussion forums and net-meetings, are not fully satisfactory. To address this issue, Internet-enabled collaborative engineering and the related techniques are developing at a rapid pace since the end of the last century. A variety of commercial tools have been launched to support engineering distribution and collaboration, e.g., SmarTeam™ (www.smarteam.com), TeamCenter™ (www.ugs.com/products/teamcenter), ProjectLink™ (www.ptc.com/products/windchill/projectlink.htm), Onespace™ (www. onespace.net), Adaptive Media Envision3D™ (www.adaptivemedia. com), Streamline™ (www.autodesk.com), etc. As a successful industrial application example, Ford worked collaboratively with its acquired Volvo from their separate sites on car design based on a collaborative design platform. With the help of collaborative tools, small and medium enterprises or even individual designers with specific domain knowledge will be able to participate and collaborate in the design process with large manufacturing companies.

An Internet-based collaborative design environment, which is illustrated in Fig. 1.4, consists of two capabilities, namely, distribution and collaboration. Distribution means systems that are geographically dispersed can be linked to support remote design activities, while collaboration allows individual designers to be associated and coordinated to fulfil a global design target and objective. A collaboration mechanism requires the specific design of a distributed architecture of a system to meet the functional and performance requirements.

CE and Internet-based collaborative engineering are complementary in functions. The former emphasises on a vertically seamless integration of the upstream design and downstream manufacturing processes through leveraging on some intelligent strategies to realise information sharing and flowing. The latter focuses on the horizontally interpersonal aspects of group work across the whole design chains. An Integrated and Collaborative Environment (ICE) for product development is actively being investigated to incorporate the two strategies to meet the following requirements:

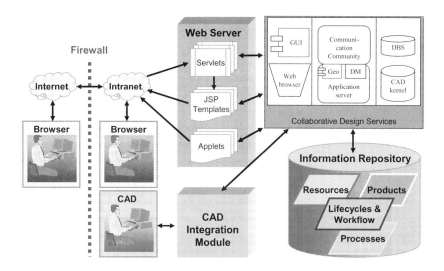

Fig. 1.4 A collaborative product development environment over the Internet/Intranet.

- Enterprise integration for the distributed organisations and systems. Manufacturing companies and enterprises can be integrated with their distributed systems and partners, such as customers and suppliers, via networks to establish an ICE for product development. For instance, the distributed design and manufacturing analysis systems can send the demands and requirements from the customers directly to the design department of a company to support global competitiveness and rapid market responsiveness.
- Heterogeneous environments and interoperability of software tools. An ICE will allow the integration and interoperability of heterogeneous software and hardware. Information environments and legacy systems in companies are usually based on different

programming languages, representation languages and models for product information, and computing platforms. To achieve an effective and efficient interoperation and interaction of sub-systems and software components in such heterogeneous environments, automatic information conversion and interpretation capabilities are necessary to realise obstacle-free information communication and workflow control.

- Open and scalable computing structure and services. There is a need to provide a possibility to dynamically integrate new sub-systems into or remove existing sub-systems from an ICE-enabled product development with high convenience, security, reliability and without stopping and re-initialising the entire environment. New kinds of service architectures to wrap software tools have to be developed to incorporate them into the environment as required, so as not to interrupt organisational links previously established.

- Cooperation between humans, and between systems and humans. People and software systems need to work at various levels of collaboration, and with rapid access to knowledge and information repositories. Bi-directional communication infrastructures are necessary to allow effective and quick communication between systems or between humans and systems to facilitate their interactions.

1.2 Enabling Technologies

Product design and manufacturing are semantically rich domains. Problems in these domains, such as reasoning, analysis, process planning, etc., are difficult to solve as a large amount of data and information are available, and complex decision-making processes are involved. It is possible to utilise AI techniques as an enabling technology to facilitate the relevant applications and to further form an integrated environment. Meanwhile, the advances of collaborative capabilities in systems is fully driven by the development and deployment of the Internet technologies, such as Java, .Net, Web, HTML (HyperText Markup Language), XML (eXtensible Markup Language) or Web service techniques, for building up the information infrastructures. In the following content, the features of these enabling technologies, especially those used in the rest of the book, will be highlighted.

1.2.1 *Artificial intelligence*

AI is the branch of computer science dealing with the design of computer algorithms and systems that exhibit characteristics associated with intelligence in human behaviours, including reasoning, learning, self-improvement, goal-seeking, self-maintenance, problem solving and adaptability. With the increase of hardware speed and advances in AI methodologies and algorithms, AI techniques have been widely used in design and manufacturing activities to solve problems in an efficient and successful way [Dagli, 1994; Teti and Kumara, 1997; Rao, *et al.*, 1999]. For instance, the Artificial Neural Network (ANN), which can be functional in pattern recognition and pattern association, has been successfully applied to manufacturing feature recognition to convert a product design model to a downstream manufacturing model. The Genetic Algorithm (GA), the Simulated Annealing (SA) algorithm and the Tabu Search (TS) algorithm can be utilised in the optimisation of process plans for products to reduce their manufacturing cost and improve the quality of the solutions.

1.2.1.1 Artificial neural network

An ANN is an information-processing technique that has certain performance characteristics in accordance with a biological neural network. An ANN, which is a specific mathematical model and algorithm to simulate neural biology, can be applied to numerous problems, such as storing data and patterns, classifying patterns, performing general mapping from input to output, and finding optimal solutions to optimisation problems [Fausett, 1994]. Usually, an ANN consists of four parts: (1) a series of simple elements, namely neurons (or units, nodes, or short-term memory elements); (2) neurons are linked to form a network and each link is associated with a weight; (3) each neuron has an input and an output, and is associated with an activation function to sum up the weighted input signals from its connected neurons to determine its output signal; and (4) each network has one input and one output, and signals are passed from neurons for input to neurons for output over connection links. A multi-layer neural network is illustrated in Fig. 1.5(a), and the computation process for a neuron y_l is shown in Fig. 1.5(b).

There are three design considerations of an ANN: (1) its architecture for organising neurons and their links, (2) a learning method to determine the weights on the links, and (3) an activation function of the neurons.

An ANN can provide a high degree of robustness due to the massive parallelism in its design. It is often used in situations where only a few decisions are required from a massive amount of data, or a complex non-linear mapping needs to be learned. Intelligent manufacturing is one of the most important and successful application domains for ANNs. Two ANNs commonly used in manufacturing feature recognition are: the Multiple-Layer Feed-Forward (MLFF) net trained using a Back-Propagation (BP) training algorithm, and the Adaptive Resonance Theory (ART) net.

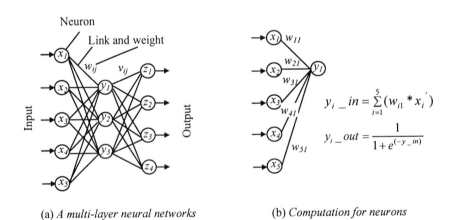

(a) *A multi-layer neural networks*   (b) *Computation for neurons*

Fig. 1.5 A multi-layer neural networks and the computation process.

(1) MLFF net with BP

An MLFF net with BP is a gradient descent method to minimise the total squared error of the output computed by the net. An MLFF net with BP is a supervised-learning net, i.e., the net can be trained to map a given vector of inputs to a specified vector of target outputs. After training, it can give reasonable or good responses to an input that is similar to that used in training. The process of computing signals in an MLFF net is forward from the input layer to the output layer, whereas the training process (BP algorithm) of the net is in reverse. This process iterates until the weights of the net are adjusted (trained) to generate the same or similar signals for the input signals. The computation processes are illustrated in Fig. 1.6.

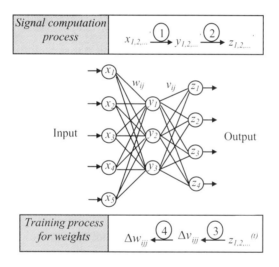

Fig. 1.6 Signal computation process and training process.

The main parameters to be determined for an MLFF with BP include the number of hidden layers, the number of nodes in the hidden layers, the choice of the initial weights and biases, and certain control parameters, such as the activation functions and the learning rate of the net. Since an MLFF net with BP can perform a stochastic gradient descent in a weighted space and can be easily realised, it has emerged as the most popular algorithm in the category of supervised training algorithms. However, there are some serious drawbacks of the net that limit its potential applications, namely:

- There is no specific mathematical method to determine the main parameters of the net, such as the number of hidden layers, the number of nodes of the hidden layer(s), etc. The determination of these parameters is quite time-consuming and tedious as it requires many trials;
- Since the mechanism of the net is a gradient descent method, results are liable to be trapped in some local minimum points. Whether a global minimum can be achieved or not depends much upon human experience and trials.

(2) ART net

An ART net is an unsupervised learning net, i.e., through adjusting the vigilance parameter in the net, the similarities among the input patterns can be controlled and these patterns can be categorised into different families.

An ART net, as shown in Fig.1.7, usually involves three groups of neurons, namely, input processing units (F_1 layer), cluster units (F_2 layer) and reset units. The F_1 layer consists of n numbers of each type of units (where n is the dimension of an input pattern). In the F_2 layer, the signals of the input units are combined and the similarities of the input vectors are compared with the weights of the cluster units. There are two sets of weights between the F_1 and F_2 layers. The bottom-up weights from F_1 to F_2 are denoted as b_{ij}, and the top-down weights from F_2 to F_1 are t_{kj}. Based on a "vigilance parameter" (ρ) in the reset units, the cluster units can decide whether to learn an input vector or not. Compared with an MLFF net with BP, an ART net has two advantages:

- An ART net has the ability to remain stable to preserve significant past learning while it remains adaptable enough to incorporate new information, whenever it might appear;
- The stability time for an ART net is much shorter than the training time of an MLFF net with BP. There are also fewer control

parameters in the ART net. Compared with an MLFF net, the ART net is easier to control and manipulate.

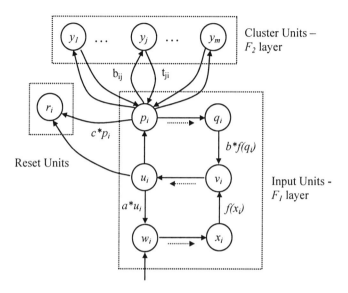

Fig. 1.7 A typical ART architecture.

There are two kinds of ART nets, namely, the ART 1 net and ART 2 net [Fausett, 1994; Haykin, 1999]. The architectures and basic mechanisms of the ART 1 and ART 2 nets are similar. However, ART 1 is designed to cluster binary input vectors and ART 2 handles continuous-valued input vectors. Therefore, an ART 2 net can be applied to more complex situations and wider application domains than an ART 1 net, benefiting from its more flexible input representation scheme. On the other hand, the input field of ART 2 is more complex since it handles continuous-valued input vectors and needs to avoid vectors that are arbitrarily close together. The architecture and computing process of an ART 2 will be explained in Chapter 3, in which a hybrid AI technique, including heuristic reasoning, graph manipulation, and an ART 2 net, has been designed for recognising interacting manufacturing features from a design part. A comparison of applying ART 2 and MLFF nets for this problem is made to show their characteristics.

1.2.1.2 Heuristic optimisation techniques

Conventional optimisation algorithms are often incapable of optimising non-linear multi-modal functions. A random search method might be helpful to search for an optimal solution. However, an undirected search is inefficient for large domains. To address this problem effectively, some heuristic optimisation techniques, such as GA, SA and TS, have been proposed to quickly find a solution in a large searching space through some embedded intelligent heuristic strategies [Pham and Karaboga, 2000]. Some basic concepts of these techniques are as follows:

- Representation. A heuristic algorithm does not use much knowledge about an optimisation problem and deals directly with the parameters of the problem. It can work with chromosomes (for GA) or solutions (for SA and TS) that represent the parameters. The optimisation problem can be represented as a string, which includes two popular schemes: binary string representation and integer/real number representation. The scheme to use is determined by the ease of modelling the problem itself as well as the performance of the algorithm in terms of accuracy and computation time.

- Evaluation. A fitness (objective) function is modelled based on certain criteria to evaluate the chromosomes or solutions. The role of the fitness function is twofold: (1) it is used as a selection operation to determine the individual chromosome or solution to be reproduced to drive the optimisation procedure; and (2) it is used as a stopping condition to determine whether a chromosome or solution has been satisfied, so as to stop or continue the search procedure. A suitable fitness function can be a mathematical equation. Where this method cannot be used, a rule-based procedure can be constructed.

- Optimisation strategies. Based on the chosen representation scheme, a GA will form a group of initial chromosomes (i.e., a population) for certain genetic operations, such as selection, crossover and mutation, to be executed on to achieve an optimal solution. For SA and TS, starting from a single solution, some neighbourhood strategies are applied to generate a group of solutions and a solution is selected according to certain criteria as a starting point to continue the optimisation process.

(1) Genetic algorithm

A GA is modelled based on natural evolution in that the operators a GA employs are inspired by a natural evolution process. These genetic operators, including selection, crossover and mutation, can be used to manipulate the chromosomes (solutions to a problem) in a population over several generations to improve its fitness function gradually. The procedure of a GA is shown in Fig. 1.8.

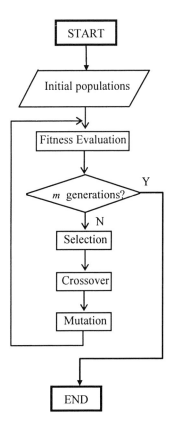

Fig. 1.8 The flowchart of a GA.

A GA works with chromosomes that represent the parameters. Several important issues to be considered when applying a GA to an application problem include:

- Representation schemes of the chromosomes, and the definition of an evaluation function.
- The selection of a suitable procedure for each genetic operator to improve the efficiency and quality of the search. For instance, for the selection operator, there are mainly two alternative procedures: proportional selection ("roulette wheel") and ranking-based selection. In the crossover operator, some common alternative strategies include the one-point crossover, two-point crossover and cycle crossover.
- In the designed chromosomes, there are usually some precedence constraints in them. The crossover and mutation operations employed in a GA might cause the precedence constraints to be violated. The method to handle constraints and search the feasible space is a major difficulty in applying GA.
- The choice of the GA control parameters depends on the problem and the representation schemes employed. Important parameters include the population size, the crossover rate and the mutation rate. These parameters should be designed for a general condition for the problem instead of being specific for a certain case study.

(2) Simulated annealing algorithm

An SA is a stochastic searching algorithm based on the principle of the real annealing process in metallurgy. It comes from an algorithmic analogy with the annealing of materials where the purpose is to lead the material to a state corresponding to a global minimum of its internal energy. This process is attained by searching the global minimum of an objective function in a space of solutions. An SA uses a control parameter called "temperature" that is decreased through iterations until it becomes close to zero [Kirkpatrick 1983]. The solution is sought by iterating and evaluating the energy at each stage. This method allows the algorithm to generate random points in which the fitness function has a greater value, i.e., the solution has a higher energy. When the temperature is high, the algorithm will be likely to accept a higher energy solution, while at a very low temperature the algorithm will almost always only accept solutions of lower energy. Solutions are accepted according to the Boltzman probability [Aarts, 1989], and new solutions are randomly chosen in a feasible domain to satisfy constraints. The temperature begins at a high level and is cooled until an equilibrium is reached by allowing the initial temperature to seek a global optimum. Without this feature, it is possible to be trapped in a local minimum. By allowing the function to move to a higher value, it is able to climb over the hill and find the global minimum, as illustrated in Fig. 1.9. Several important issues to be considered when applying a SA include:

- Representation schemes of the solutions.
- Definition of an evaluation function.
- Definition of a neighbourhood mechanism for the generation of temporary solutions. Various neighbourhood mechanisms for the generation of temporary solutions can be developed while some of them could be adopted from a GA, for instance, crossover and mutation operators.
- Design of a cooling schedule. The parameters in the cooling schedule to be determined include an initial temperature, a temperature update rule, the number of iterations to be performed at each temperature step and a stopping criterion for the search.

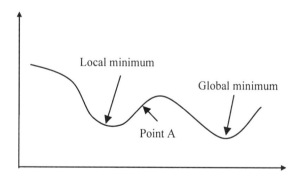

Fig. 1.9 Hill-climbing analogy in an SA.

It has been recognised that GA is not well-suited to perform finely-tuned local search [Rogers, 1991; Ishibuchi, *et al.*, 1994; Mathias, *et al.*, 1994; Yen, *et al.*, 1998]. Compared with a GA, an SA is more efficient in searching for a global or near-global optimisation solution, but inefficient in obtaining a group of solutions with good performance. Meanwhile, it has been shown that to control and adjust the parameters of an SA in an entire search space is not easy. Therefore, in the case of searching for a group of solutions for a global or near-global optimisation solution, it is meaningful to make an attempt to combine the strengths of a GA and an SA. Once the high performance regions of a search space have been identified using a GA, an SA should be invoked as a local search routine to optimise the members of the final population. In Chapter 5, a hybrid GA-SA method has been designed to optimise a process planning problem for optimal or near-optimal solutions.

(3) Tabu search algorithm

TS, which utilises some selected concepts of GA and SA and is characterised by the use of a flexible memory strategy, is a meta-heuristic algorithm that can guide search processes to overcome local optimal solutions in combinatorial optimisation problems. The fundamental of TS is to avoid entrapping in cycles by forbidding moves that take the solution to points in the solution space previously visited (hence "taboo") [Glover, 1997]. Although TS is still in its infancy stage, during the last few years, it has been reported as a satisfactory solution approach for a variety of problems, such as scheduling, parallel computing, transportation, routing and network design.

In a TS algorithm, there are three main strategies, namely, the forbidding strategy, the freeing strategy and the aspiration strategy. During a search process, a Tabu list for recording the recently past moves is established and dynamically maintained. The strategies are briefly explained below.

- The forbidding strategy controls the solution that enters the Tabu list. It can avoid cycling and local minimums by forbidding certain moves during the most recent computational iterations.

- The freeing strategy is used to manage the solution that exits the Tabu list and when this occurs.
- The aspiration strategy is the interplay between the forbidding and freeing strategies for selecting trial solutions. A solution that has been forbidden by the Tabu list can become acceptable if it satisfies a certain criterion.

In Chapter 7, a TS-based approach is used to optimise a process planning problem for optimal or near-optimal solutions, and the relevant results are compared with the GA and SA approaches.

1.2.2 *Internet technologies*

The Internet technologies are evolving quickly recently. Many competing or complementary technologies have been launched by large software vendors, such as Microsoft, Sun and IBM. A few popular technologies are briefly introduced here, and information for applying the relevant Internet technologies to establish a collaborative system is surveyed in Chapter 6.

1.2.2.1 Client/server architecture

In Internet applications, the client/server architecture is the most commonly used paradigm to separate business logic, data processing and presentation of data. Clients are graphical user interfaces to present and render data, and servers serve as data repositories or executive processors for the data according to certain logics and programs. The original architecture of a client/server system is two-tier, and later migrates to three- or even n-tier (a three-tier architecture is shown in Fig. 1.10).

The motivation behind an n-tier architecture is to separate the application logic (in servers) from the user interface (in clients), which brings flexibility to the design of an Internet application. For instance, multiple user interfaces can be built without changes to the application logic and the relevant programs. Meanwhile, administration functions can be deployed in a server for effectively coordinating and monitoring the clients as a team. Recently, there have been much research activities on a peer-to-peer architecture to allow a group of computers with equivalent responsibilities to be connected so as to pool their resources and decentralise the management, which is suitable for multi-agent systems, parallel computing and grid computing. To choose a suitable architecture and arrange the functional components in different computers in a network requires careful investigations according to the practical requirements and conditions.

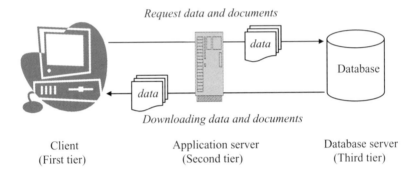

Request data and documents

data

Database

data

Downloading data and documents

Client	Application server	Database server
(First tier)	(Second tier)	(Third tier)

Fig. 1.10 A three-tier client/server architecture.

1.2.2.2 Representations for documents and design models

Along with HTTP (HyperText Transfer Protocol), which enables the cross-platform and cross-enterprise multi-media exchange of information from a server to a client or an end-user, HTML is a scripting language to organise texts, pictures, data and documents and render them in a Web browser in certain formats. HTML information is saved in plain text (e.g., ASCII), and the representation formats of the data and information are controlled by the HTML commands ("tags"), attributes and values. A drawback of HTML is that it does not allow users to create their own tags or attributes to parameterise or semantically specify their data. Another drawback is that it lacks a complex structure definition such that it cannot support the specification of structures needed to present database schemas or object-oriented hierarchies. Hence, design and manufacturing information loses its structure when translated into a HTML file, and the embodied information is difficult to process and be extended.

To eliminate the drawbacks of HTML, XML was created to support distributed structured data and documents over the Web. XML offers several potential advantages that can be used in a collaborative environment. For instance, XML can define data in a document-oriented presentation format to separate a document's logical structure from its document-oriented presentation. Based on this characteristic, a user can specify several styles for the same XML document, and the same document structure can be defined as several output formats for different applications in the design and manufacture domains. XML can build documents from heterogeneous data and information. Hence, design and manufacturing applications, which might include heterogeneous information sources, such as documents, knowledge bases, relational or object-oriented databases, and case bases, can be integrated through utilising XML documents.

Research on the sharing of multi-media information, especially 3D modelling geometrical information on the Internet/intranet in real-time has been carried out recently. Such a capability can support heterogeneous software tools in an ICE, shorten a product development process as well as improve its quality. VRML (Virtual Reality Mark-up Language) is a language for describing interactive simulation of 3D models to be included in a HTML file for rendering in a Web browser. Essentially, VRML files describe a 3D scene in terms of objects, operations, and properties of the scene. Triangle, quadrangle and hexagon are several popular schemes to represent the meshes of a VRML model. Fig. 1.11 shows an example of the VRML syntax and its corresponding visual effect as seen in a VRML viewer. The basic geometry information is represented by the contents in *point* (providing the 3D coordinates of vertices) and *coordIndex* (providing the vertex indices in a certain sequence to compose the triangular meshes).

Other standards, such as X3D (eXtensible 3D) (www.x3d.com) and MPEG-4, which are functionally equivalent to VRML, extend the support of XML-based representation and video/audio application in compressed binary formats, respectively. OpenHSF (www.openhsf.org)

and XGL/ZGL (www.xglspec.org) enhance the capability of VRML for effective 3D streaming transmission of large-volume data over the Internet through data compression, mesh simplification and object prioritising to facilitate real-time collaboration.

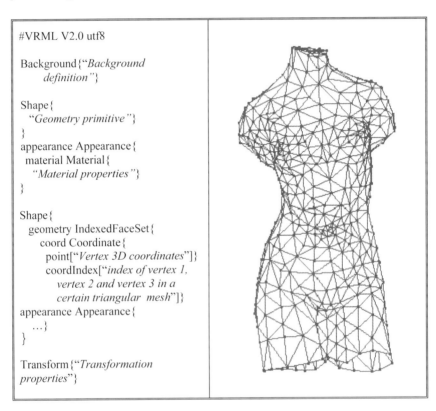

```
#VRML V2.0 utf8

Background{"Background
    definition"}

Shape{
    "Geometry primitive"}
}
appearance Appearance{
 material Material{
    "Material properties"}
}

Shape{
  geometry IndexedFaceSet{
     coord Coordinate{
        point["Vertex 3D coordinates"]}
        coordIndex["index of vertex 1,
           vertex 2 and vertex 3 in a
           certain triangular  mesh"]}
  appearance Appearance{
     ...}
}

Transform{"Transformation
properties"}
```

Fig. 1.11 The VRML syntax and its corresponding visual object.

1.2.2.3 Distributed enterprise system integration paradigms

Microsoft's .Net, OMG (Object Management Group)'s CORBA (Common Object Request Broker Architecture) and Sun Microsystems' J2EE (Java 2 platform Enterprise Edition) are the most popular distributed enterprise system integration paradigms used to establish a more complex collaborative product development environment on the Internet/Intranet. Each strategy has its characteristics. .Net is now heavily used on the Microsoft Windows platform, while CORBA and J2EE can be used on diverse operating system platforms from mainframes to UNIX boxes to Windows machines.

J2EE is chosen as the implementation paradigm to realise two prototype systems in Chapters 7 and 8. They are namely a Web-based prototype system for users to carry out visualisation-based design and manufacturing analysis, and a distributed co-design prototype system based on a J2EE infrastructure to enable a dispersed team to accomplish a collaborative feature-based design task, respectively. The underlying techniques for the J2EE paradigm, especially those used in this book, are briefly introduced here.

The J2EE, CORBA and .Net paradigms can be used to:

- Encapsulate and integrate existing legacy systems in product design, planning, simulation, execution, and distribution into an open, distributed intelligent environment via networks.
- Provide interfaces for designers and software systems to realise human/system interactions and interoperability of systems.
- Model special services in a distributed application, such as registration service, administration services, and communication facilitator, to coordinate product development processes and manage information and data.

A J2EE paradigm is composed of a series of disparate technologies that can be categorised as the component technologies, service technologies and communication technologies [Allamaraju, *et al.*, 2001].

Servlet and JSP (JavaServer Pages) are two Web-based component technologies that provide the server-side programming means to extend the functionality of a Web server to provide dynamic contents in HTML or XML. This is illustrated in Fig. 1.12.

Service technologies include the JDBC (Java DataBase Connectivity), JMS (Java Message Service), JCA (Java Connector Architecture), etc. These services are used to manage the access of databases, send and receive messages, integrate legacy applications, etc.

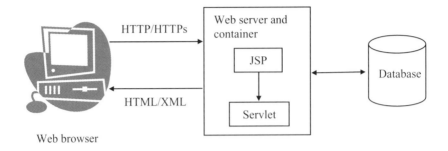

Fig. 1.12 Servlet- and JSP-based system architecture and the communication.

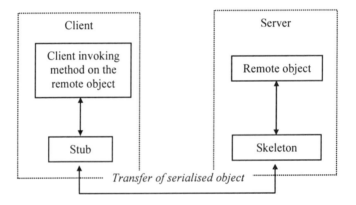

Fig. 1.13 A communication process based on RMI.

Communication components include the fundamental Internet protocols, such as HTTP and TCP/IP, and remote object protocols, such as RMI (Remote Method Invocation) and RMI-IIOP (RMI over Inter-ORB Protocol). TCP and IP are two fundamental transmission and Internet protocols, and they work together as a single entity to move data around the Internet. HTTP is a generic, stateless and application-level protocol that enables the multi-media exchange of information from a server to a client. Based on TCP/IP, RMI, which is one of the primary mechanisms in distributed object applications, allows object-to-object communications between different systems. Based on RMI, applications can be enabled to call object methods located remotely and share data across systems. RMI defines interfaces of remote objects, and methods on these remote objects can be called through stub and skeleton mechanisms as if they are local. A RMI process is illustrated in Fig. 1.13. RMI-IIOP extends RMI to support CORBA mapping and invocation. With RMI-IIOP, a remote interface to any remote object implemented in any language can be defined and invocated.

1.3 Summary

With the extensive global competition and the rapid evolution of the Internet technologies, product development systems are moving towards supporting integrated and collaborative applications. ICE systems have been actively investigated to provide solutions to integrate geographically dispersed users, systems, resources and services, and enable them to cooperate and interoperate in an Internet/Intranet environment beyond the physical and temporal boundaries. This chapter presents an introduction of the motivations, concepts and objectives of establishing an ICE for product development. Some enabling technologies and their features, especially those used for system and methodology implementation in this book, are highlighted.

Chapter 2

Manufacturing Feature Recognition Technology – State-of-the-Art

In a CE environment, downstream manufacturability activities are considered simultaneously during the design phase of a part. The design task of a part is usually carried out in a Computer-Aided Design (CAD) system, in which geometric functionalities are provided for embodying design intent and visualising the part. Recognition of manufacturing features from the design part, which is an important technology in a CE environment to seamlessly and efficiently link a CAD system and the downstream manufacturing applications, such as CAPP (Computer-Aided Process Planning) and CAM (Computer-Aided Manufacturing), can associate certain geometric and topological patterns in the part with manufacturing semantics in terms of manufacturing features.

In this chapter, the manufacturing feature recognition technology is first introduced under the retrospection of the development of CAD systems and representation strategies for design models. Some previous works in the manufacturing feature recognition technology are categorised and the main characteristics of each category are summarised. Finally, this chapter is concluded and the future research directions are highlighted.

2.1 Evolving Representations for Design Models

During the past several decades, CAD has experienced major paradigm shifts and technological innovations. Three events can be identified as the milestones in the historical development of the CAD industry, namely,

- The introduction of 2D drafting systems;
- The introduction of 3D modelling systems; and
- The introduction of feature-based systems.

2D drafting systems have been developed to provide designers with a convenient way to represent 2D drawings based on a variety of 2D geometric elements and annotation methods to define product shapes and tolerances. The earliest academic 2D system was the Sketchpad system developed in the mid-1960s at MIT [Sutherland, 1963]. In 1970s and 1980s, companies in the automotive and aerospace industries launched several commercial 2D drafting systems, which include the early versions of ADAM™ from MCS Inc. and AutoCAD™ from AutoDesk Inc. However, these systems have some limitations in product design. For instance, for complex product models, skills are needed to construct and interpret 2D drawings, which are error-prone and time-consuming.

From the mid-1970s, surface modelling and solid modelling techniques, which form the foundations of the 3D geometric modelling systems, were developed and they provided a more effective way to represent models of complex components that are difficult to be described and interpreted in 2D drawings. Major CAD systems, such as Unigraphics™, Pro/E™, I-DEAS™, CATIA™, Inventor™, have been developed to offer 3D geometric modelling functions.

In surface modelling, well-known schemes, such as Coons [Coons, 1964], Bezier [Bezier, 1966], B-Spline [Schoenberg, 1946; Cox, 1972; de Boor, 1972; Gordon and Riesenfeld, 1974] and NURBS (Non-Uniform Rational B-Spline) [Versprille, 1975], were introduced and successfully applied for product design in shipbuilding, automotive, aircraft, mould and die industries. These schemes approximately represent curves and surfaces such that they can be manipulated through adjusting control points and weights. Rational B-Splines and NURBS are the two major schemes for representing curves and surfaces. In a surface model, however, there is no information to distinguish which part of a surface model is solid, and the topological connections between surfaces for constructing a complete 3D model are lacking.

The aim of the introduction of the solid modelling technique in 1970s was to generate unambiguous and complete 3D models. Two major schemes, which have been widely applied in most of 3D geometric modelling systems, are CSG (Constructive Solid Geometry) and B-Rep (Boundary Representation).

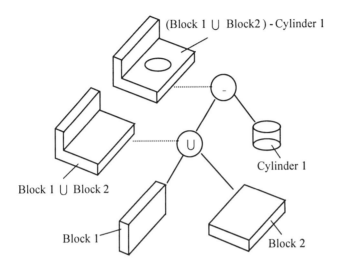

Fig. 2.1 A CSG tree.

CSG, pioneered by Voelker and Requicha [1977], is a binary tree structure for creating 3D models based on some primitive shapes and three Boolean operations, namely, union ($\cup$), difference (-) and intersection ($\cap$). Fig. 2.1 illustrates the structure of a CSG tree and its corresponding geometry. In a CSG tree, primitives, which are leaves and associated with dimensions, positions and orientations, are building blocks, and Boolean operations, which are intermediate nodes, are used to combine primitives. A CSG tree is a compact representation that can explicitly reflect the construction history of a product model. It is not unique, however, as there are many different ways in which primitives, transformations and operations can yield the same part. Another drawback of a CSG tree is that it is an implicit form for its geometric elements (e.g., vertices and edges), and extra computation burdens, which are caused by applying Boolean operations on the primitives of the CSG tree, are necessary to evaluate these elements.

Based on surface models, B-Rep is a data structure that explicitly represents the topological relationships of some geometric elements, such as vertices, edges, loops, surfaces and compounds, to form a solid. The faces and edges in a B-Rep have compact mathematical representation. For instance, faces are represented as planar, quadratic, toroidal, or parametric surfaces, and edges can be represented in parametric forms. The winged-edge data structure [Baumgart, 1974] and the half-edge data structure [Mantyla, 1988] are two popular B-Rep implementation schemes. The winged-edge data structure, which is based on the observation that every edge has exactly two "next" edges and two "previous" edges, is a common data structure to represent the B-Rep of a manifold model. An example of a B-Rep winged-edge data structure is shown in Fig. 2.2.

The historical developments of CAD, CAPP and CAM are quite independent of each other. This has caused some drawbacks in the integration of product design and manufacturing processes, namely,

- A designer, a process planner or a machinist has different viewpoints of a design part. For instance, the CSG model of a part reflects a constructive process and the geometric primitives can be additive volumes. From the manufacturing viewpoint of the part, the geometric entities that are associated with machining processes should be subtractive in volumes to reflect an incremental decomposition process. B-Rep is also not a suitable format for manufacturing analysis as its information is organised as some low-level geometric entities, such as vertices, edges, loops, faces, etc., which are not easily associated with manufacturing processes. For a part, a process planner or machinist needs to re-interpret the design information from the manufacturing viewpoint, which might be time-consuming and causes human error.

- An alternative approach to ease or bypass the above difficulty is to design products in terms of manufacturing entities. However, this approach blurs the functional differences among design, analysis and manufacturing, and forces a designer to think in a way that is not natural to him or her.

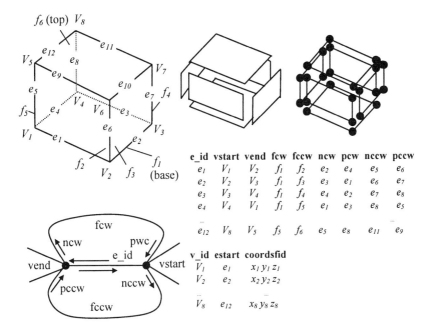

Fig. 2.2 A winged-edge data structure.

To establish a more effective integrated design and manufacturing environment, the concept of features came into existence in 1980s to address the above deficiencies. Features associate some generic solid shapes with engineering semantics to facilitate design and manufacturing processes. Features, which are domain-dependent and diversified in functions, are generally classified as design features and manufacturing features. Design features contribute to the design construct basis of a part in terms of additive or subtractive volumes, and their semantics reflect the design intent and function. Associated with design features is the design-by-feature approach, in which a design feature tree is used to organise the features of a part in a hierarchical structure according to the evolving design process. The root of the tree is the construct basis of the part. Each intermediate node in the tree is a Boolean union, intersection or difference operator, and each leaf of the tree is either an additive design feature volumetrically added onto the part, or a subtractive design feature volumetrically removed from the part. Each feature is associated with one or more datum planes or axes, surface tolerances, attributes, dimensions, etc. Different from a conventional CSG tree, a design feature tree defines more enriched and complex types, inter-relationships and operations of features to facilitate a product design process. In addition, features have more explicit engineering meanings for the designers, such as holes, slots, pockets, ribs, etc. Based on the design-by-feature approach, more advanced technologies, such as semantic design, constraint-based design and variant design have been actively investigated.

Manufacturing features are normally derived from a manufacturing stock in terms of subtractive volumes, and they are conveniently associated with manufacturability analysis and process planning activities, such as fixture planning, machines and cutting tools selection, and machining operations planning. STEP AP 224 is an ISO (International Standards Organisation) STEP application protocol that specifies manufacturing information and process plans using manufacturing features to machine discrete mechanical parts (ISO 10303-224, 1996). In the AP 224, four groups of manufacturing features are defined, namely, machining features, transition features, replicate features and compound features. According to the definitions of the AP 224, machining features are essential manufacturing features and consist of sixteen different categories. Transition features, which include fillet, round edge, and chamfer, are auxiliary parts of machining features or connective parts between machining features. Replicate features are a group of machining features that are copied from a single template and arranged in three patterns, i.e., circular, rectangular and general patterns. Compound features are complex feature definitions that unite several relevant machining features together. Three essential elements are required to machine the effective volume of a manufacturing feature, namely, the Tool Approach Direction (TAD, a vector), the cutting tool position face, and the cutting depth. They can be expressed as a triple - (TAD, tool position face, and cutting depth). The classification of machining features and a few examples of manufacturing features are shown in Fig. 2.3.

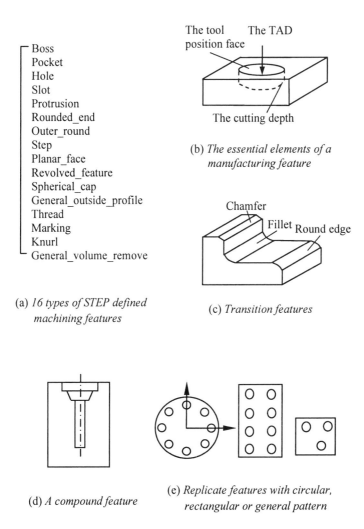

Boss
Pocket
Hole
Slot
Protrusion
Rounded_end
Outer_round
Step
Planar_face
Revolved_feature
Spherical_cap
General_outside_profile
Thread
Marking
Knurl
General_volume_remove

(a) *16 types of STEP defined machining features*

The tool position face   The TAD

The cutting depth

(b) *The essential elements of a manufacturing feature*

Chamfer   Fillet Round edge

(c) *Transition features*

(d) *A compound feature*

(e) *Replicate features with circular, rectangular or general pattern*

Fig. 2.3 STEP AP 224 defined manufacturing features.

A recognition algorithm, which works on a geometric model or a design-by-feature model to identify manufacturing features, can establish a seamless and intelligent interface between product design and manufacturing applications. Manufacturing feature recognition has been the focus of much research over the past twenty years. However, a fundamental difficulty to apply this technique in practical applications is the representation and recognition of interacting features from complex design parts. The ability to handle interactions has become a benchmark for the developed feature recognition algorithms and systems.

Depending on the feature representations and strategies adopted, manufacturing feature recognition systems can be generally categorised as the boundary and the volumetric feature recognition schemes based on the geometric modelling representations, and the hybrid scheme of the feature recognition approach integrated with the design-by-feature approach. The boundary-based scheme is based on the B-Rep or some variant formats of a design model. In the scheme, the main operation is to detect and extract manufacturing features from some typical configurations of the faces, loops, edges or vertices of a part. In the volumetric-based scheme, the input design model can be either a boundary representation or a volumetric representation. This scheme decomposes a design model as a set of features based on volumetric operations. The hybrid scheme, which integrates design-by-feature models and feature recognition processes, is a recent promising trend to use the design feature information to guide and facilitate the recognition process and alleviate the searching difficulty.

Some details of the previous works are discussed next.

2.2 Boundary Feature Recognition Scheme

The boundary scheme, which can be roughly categorised into the rule-based, the graph-based, the hint-based, and the artificial neural networks-based approaches, recognises and represents manufacturing features using the boundary information of a design part.

2.2.1 *Rule-based approach*

In the rule-based approach [Kyprianou, 1980; Henderson, 1984; Choi, *et al.*, 1984], a set of heuristic rules, coded in Prolog, LISP or some other programming languages, is defined to describe some typical template patterns of features. The entities expressed in a rule are the boundary elements in a part, i.e., faces, loops, edges and vertices. In these rules, some characteristic relationships or conditions between entities, such as parallel, perpendicular, adjacency, equality, concavity and convexity, are defined to specify some particular patterns and constraints of features. For instance, three rules used to define slot features as illustrated in Fig. 2.4 are listed below.

- Rule 1: A slot is composed of three faces $\{F_1, F_2, F_3\}$;
- Rule 2: F_1 is adjacent to F_2 and F_2 is adjacent to F_3; and
- Rule 3: F_1 forms a concave angle with F_2 and F_2 forms a concave angle with F_3.

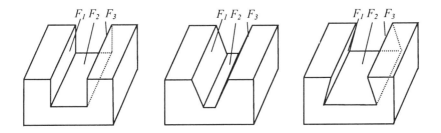

Fig. 2.4 Three slots satisfying the three rules.

In order to describe more complex situations, two additional rules need to be added as below:

- Rule 4: The degree difference between two outward normals of F_1 and F_3 is within a certain range (e.g., in Fig. 2.5(a), although F_1, F_2 and F_3 satisfy Rules 1-3, the angle difference between the normals of F_1 and F_3 is beyond the range such that they cannot form a valid slot); and
- Rule 5: The distance between F_1 and F_3 should not be too large (e.g., in Fig. 2.5(b), F_1 and F_3 are too far apart and they are not regarded as a slot in some situations).

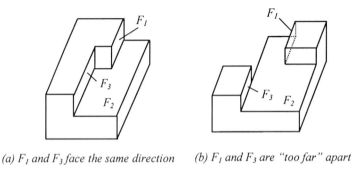

(a) F_1 and F_3 face the same direction   (b) F_1 and F_3 are "too far" apart

Fig. 2.5 Two invalid slots.

Based on the rules of defining features, pattern-matching algorithms and expert systems are employed to search for features in a design part. The advantage of the rule-based approach is that it provides an easier way for process planners and analysts to describe the features of interest. The shortcomings include: (1) it is difficult to define sufficient and exact rules to describe all the possible features, and ensure that these rules are mutually exclusive, especially in some complex cases where there are interacting features; and (2) recognition involves exhaustive searching processes and pattern matches from the geometric representation of a design part when it is complex.

For instance, in Fig. 2.6(a), with the interactions between *Feature A* and *Feature B*, the base face (F_2) of *Feature A* is removed. Based on the rules to define a slot feature, *Feature A* cannot be recognised. In order to handle this case, only Rule 4 can be used for identifying this kind of slot. Another example is shown in Fig. 2.6(b). With the addition of a new geometry (*Feature B*) that interacts with some existing entities (*Feature A*) in a design part, a face might be divided into several isolated faces and the detection of features is hindered. For Rule 5, it is vague to define a distance to be "too far". It is not appropriate to give an exact definition or upper bound to clarify the vagueness since different workshops have different manufacturing capabilities. Therefore, the rule-based approach cannot ensure the adaptability of recognition systems in handling various interacting features.

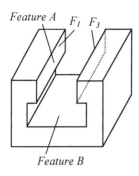 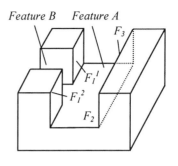

(a) The base face (F₂) of
Feature A is lost

(b) F₁ is split into two
separated faces, F₁¹ and F₁²

Fig. 2.6 Changes of geometrical and topological relationships in features due to feature interactions.

2.2.2 *Graph-based approach*

The graph-based approach [Joshi and Chang, 1988; DeFloriani, 1989; Falcidieno and Giannini, 1989; Sakurai and Gossard, 1990; Gayanker and Henderson, 1990; Chuang, 1991; Corney and Clark, 1991; Fields and Anderson, 1993; Venuvinod and Wong, 1995; Senthil kumar, *et al.*, 1996; Tuttle, *et al.*, 1998; Ibrahim and McCormack, 2002] reduces the search space by organising some important boundary entities and the relevant information of a design part into graph structures with attributes. This approach can facilitate the recognition of features with identical topology but differing geometry. The graph representation can naturally describe the simplified B-Rep of a design part in a concise way. Graph manipulation and matching algorithms can be applied to extract features.

Two types of graph representations are commonly used, namely, the face-edge graph and the edge-vertex graph. Comparing these two representations, the face-edge graph is a less ambiguous representation of a solid model [Gayankar and Henderson, 1990]. The popularity of the face-edge graph depends upon the following observations:

- Faces in a part are the distinct geometric elements to reveal the possible utilised machines and cutting tools since faces can hold the cutting tools and guide/constrain the movements and operations of the cutting tools;
- The convexity of the connecting edges between faces can indicate the existence of manufacturing features. For instance, a concave edge between two adjacent faces usually means that a manufacturing volume removal is needed on a stock to generate this shape, and the removed volume can be related to manufacturing features; and
- From the topological relationships of faces and edges, manufacturing features can be classified.

The Attributed Adjacency Graph (AAG) proposed by Joshi and Chang [1988] is a typical face-edge graph, in which the nodes of the graph represent the faces in a part, and the arcs with attributed values represent the edges between two adjacent nodes (faces). The value of each arc depends upon the connection type (convexity or concavity) of the edges. For instance, in Fig. 2.7, a part and its AAG are shown. Based on graph manipulation algorithms, an AAG can be decomposed into subgraphs and matched to some feature categories. DeFloriani [1989] developed a varied face-edge graph called the Generalised Edge-Face Graph (GEFG) to incorporate edge loop information to facilitate the recognition of features. Certain types of features, such as protrusions, depressions, through holes, handles and bridges nest on some faces of a part and generate "inner" edge loops (Fig. 2.8).

In order to handle interacting features, some heuristic rules, procedures or hypotheses have been developed for restoring the original relationships between the entities in features that have been distorted by the interactions. For instance, two types of feature interactions identified by Joshi and Chang [1988] are listed below and illustrated in Fig. 2.9:

- Type 1: Features interact such that they only have common edges between them, which cause the faces of one feature to be split up (e.g., $F_1^{\ 1}$ and $F_1^{\ 2}$ in Fig. 2.9(a)). Face pairs are unifiable if they satisfy the following two conditions:
  (1) The two faces have the same equation; and
  (2) The two faces have one face to which both are adjacent to.

- Type 2: Features intersect such that they share a common face, and interaction between the features splits a face of the feature (e.g., the shared face F_2 belonging to two slots in Fig. 2.9(b)).

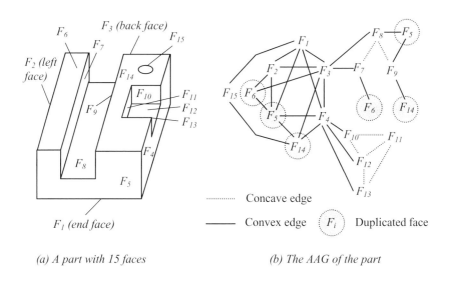

(a) A part with 15 faces   *(b) The AAG of the part*

Fig. 2.7 A part with its AAG representation.

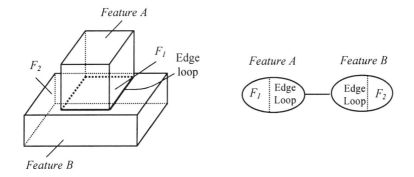

Fig. 2.8 An edge loop between two features.

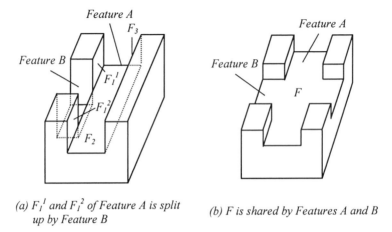

(a) $F_1{}^1$ and $F_1{}^2$ of Feature A is split
    up by Feature B

(b) F is shared by Features A and B

Fig. 2.9 Two types of feature interactions.

For these two kinds of interactions, several graph procedures have been developed to restore the original topological relationship of the features for identification. In practical situations, however, sometimes the Type 1 interaction is deficient or invalid. For instance, for its condition (1), if two faces have the same equation but their normals are reversed, it is inappropriate to unify them as a face. For its condition (2), in many cases, it is unnecessary for two faces, which have the potential to be unified, to share the same adjacent face. In order to give a more accurate description, Marefat and Kashyap [1990] introduced a hypothesis to refine the conditions defined in the Type 1 interaction. The hypothesis is described as below.

Hypothesis: a set of faces $U = \{F_1, F_2, ...F_i\}$ is unifiable if:
- Every face in U is embedded on the same plane, i.e., they have the same equation;
- No face of the part intersects the interior of the face $\{F_1, F_2, ...F_i\}$ formed by unifying $F_1, F_2, ...F_i$; and
- Normals of all the faces in U are in the same direction.

In order to restore other kinds of connected faces that have been destroyed through feature interactions, a "virtual link" concept was introduced by Marefat and Kahsyap [1990]. A virtual link is a lost concave edge between two orthogonal faces and it can be established by extending these two faces to intersect each other. An example of two virtual links is shown in Fig. 2.10. However, in practical situations, not all virtual links are valid connections in parts. Marefat and Kahsyap [1990] applied uncertainty theories to evaluate some augmented evidences of the virtual links to determine the actual connected faces. Trika and Kashyap [1994] and Gao and Shah [1998] extended this work in the generation of virtual links and the verification of feature hypothesis based on geometric reasoning.

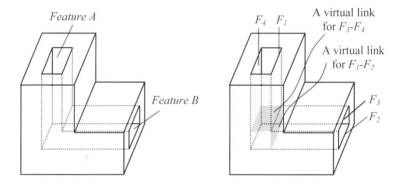

(a) Two features interact in a part   *(b) Two virtual links between two sets of faces*

Fig. 2.10 Two virtual links to restore features in a part.

In the graph-based approach, well-established techniques of graph algorithms can be readily applied to feature recognition. The main difficulty of the graph-based approach is similar to that of the rule-based approach, i.e., it is difficult to handle certain types of interacting and inexact features although some heuristic rules or hypotheses have been developed to address this problem. Another shortcoming of the graph-based approach is that the computing time increases rapidly for complex parts since graph matching is an exact process with the pre-defined single feature templates. However, the graph representation, which is a natural way to map the boundary information of a part, can be used for other approaches, such as the hint-based approach and the artificial neural networks-based approach, or some hybrid approaches [Zhang, *et al.*, 1998; Gao and Shah, 1998; Wong and Lam, 2000]. In Table 2.1, the related works for the graph-based approach are summarised.

2.2.3 *Hint-based approach*

The hint-based approach addresses the difficulty of recognising interacting features through simulating the human's intuitive process for observing interacting features [Vandenbrande and Requicha, 1993; Regli, 1995; Sommerville, *et al.*, 1995; Regli, *et al.*, 1997; Han and Requicha, 1998(a); Han, *et al.*, 1998]. From some characterised geometric patterns left in the nominal geometry, which are named as the hints or clues of features, some prospective manufacturing features can be detected. From the hints found, geometric completion procedures perform extensive geometric reasoning to construct the features.

In [Vandenbrande and Requicha, 1993], the concept of feature hints was motivated based on an observation that topological relationships are most easily affected by feature interactions while geometric relationships usually remain intact. In other words, a feature might leave a trace in a part boundary even when features intersect. Feature hints are defined as some incomplete patterns present in a design part that can be associated with features. Han and Requicha [1998(a)] defined hints for three kinds of features, namely, holes, slots and pockets, as follows.

- Hint for a hole. The machining operation for a hole leaves at least a face in the final part: a whole or partial cylindrical wall face. An example is shown in Fig. 2.11.
- Hints for a slot. The trace of a slot left in a part boundary can be a pair of wall faces, or a floor face and a pair of wall faces. Therefore, there are two hints for the presence of a slot:
  (1) Two faces are potential wall faces of a slot if they overlap when one is projected onto the other and their normals are opposite. In Fig. 2.12, F_1 and F_2 are two wall faces for a slot.

(2) A face is a potential floor face of a slot if it satisfies two conditions: (a) two wall faces must overlap in a region above the floor face when one wall is projected onto the other; and (b) the extent of the floor and walls in the sweep direction must all overlap. In Fig. 2.13, two wall faces F_1 and F_3 overlap above F_2, and these three faces can be extended along the same sweep direction. Therefore, F_2 is a floor face to form a slot with F_1 and F_3 as two wall faces.

- Hints for a pocket. The trace of a pocket left in a part boundary can be a set of floor faces, a set of wall faces, or both. Therefore, two kinds of geometric entities can be regarded as the hints of a pocket: namely, wall faces and floor faces. Fig. 2.14 shows a set of wall faces for two pockets.

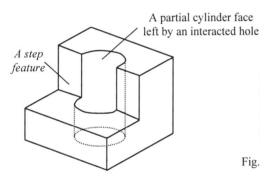

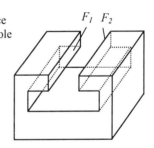

Fig. 2.12 F_1 and F_2 form wall faces of a potential slot.

Fig. 2.11 A hint for a hole.

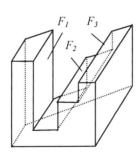

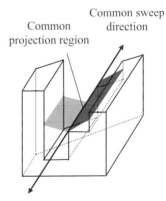

Fig. 2.13 F_2 is the floor face of a feature consisting of F_1, F_2 and F_3.

Table 2.1 A summary of some graph-based methods.

Works	Graphs	Major Characteristics	Recognised Features Types
Joshi and Chang, 1988	AAG (a face-edge graph)	Can handle two types of feature interactions: (1) features share a common edge; and (2) features share a common face or a face of a feature is split by the other.	Blind and open step/slot/pocket
DeFloriani, 1989	GEFG (a face-edge graph)	Some features leave "inner" edge loops in a design part. GEFG represents the loops and from them features are detected and recognised.	Protrusion, depression, through hole, handle and bridge
Marefat and Kashyap, 1990; Trika and Kayshyap, 1994; Ji and Marefat, 1995	Cavity graph with virtual links (a face-edge graph)	In order to restore the destroyed topological relationships by interacting features, evidences are analysed using uncertain reasoning methods to establish virtual links in the graph.	Blind and through step/slot/pocket/hole
Gayankar and Henderson, 1990	A face-edge graph	Features are recognised from edge loop features left on faces.	Protrusion and depression
Chuang and Henderson, 1990	A vertex-edge graph	Concavity and convexity of vertices are labelled to indicate the topologies of adjacent edges and faces. However, this graph is only suitable for singly connected faces due to its ambiguity.	Pockets, through hole and protrusion

Table 2.1 A summary of some graph-based methods (cont'd).

Works	Graphs	Major Characteristics	Recognised Features Types
Fields and Anderson, 1993	OFAG (a face-edge graph)	An edge is specified by two attributes according to the boundary conditions of its two adjacent faces. Merge operations are executed to combine faces to form features.	Blind and open step/slot, depression, passage and protrusion
Venuvinod and Wong, 1995	MAAG(a face-edge graph)	More detailed attributes for edges are defined such as sharp and smooth concavity/convexity.	Blind and open step/slot/pocket
Zhang, *et al.*, 1998	RAAG (a face-edge graph)	Edges are defined using convexity, concavity and virtual links. Faces to have no-convex edges indicate the presence of features and they are defined as reference faces. Features are deduced from edge attributes and reference faces.	Blind and through step/slot/pocket/hole, and protrusion
Gao and Shah, 1998	EAAG (a face-edge graph)	From the enhanced graph with virtual links insides, five types of interacting features are handled and recognised.	Blind and open step/slot depression, passage and protrusion
Wong and Lam, 2000	FES(a face-edge graph)	In the enhanced face-edge graph, the sequence of the edges of a face is defined for feature identification	Pockets, through hole and protrusion
McCormack and Ibrahim, 2002	AAG(a face-edge face)	Edge connection sequence is defined in the graph. Edge loops are used to find features.	Depression, protrusion, passage and multi-part surface

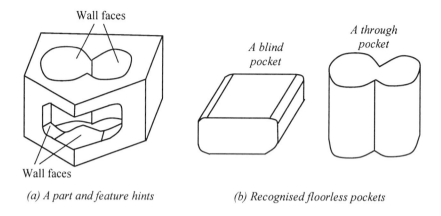

(a) A part and feature hints   *(b) Recognised floorless pockets*

Fig. 2.14 Wall faces as feature hints for pockets.

Comparing the hints for a slot and the rules defined for a slot in Section 2.2.1, it can be found that the hints of a feature are a sub-set of the rules to define the feature. In other words, the hint-based approach lifts some restrictions of the exact definitions of features and the matching processes in the rule-based approach so as to facilitate the recognition of interacting features. The complete information of features can be rebuilt based on a "generate-test-repair" paradigm. That is, generated hints need to be tested and repaired as features. Vandenbrande and Requicha [1993] and Han and Requicha [1998(a)] employed an accessibility criterion as a set of directions along which the cutter/machine assembly can move without colliding with a part, to test the validity of a potential feature with hints. The repair strategy involves a volumetric growing process, which will be discussed in Section 2.3.2.

2.2.4 *Artificial neural networks-based approach*

For different manufacturing applications, it is necessary to be able to define specific manufacturing features with various semantics. These features should be application-specific, both geometrically and topologically. Therefore, the adaptability and extensibility of recognition algorithms in handling various features are important, and this has not been effectively addressed by the above approaches.

The ANN-based approach shows a promising ability to solve the interacting features and recognise new types of features benefiting from the high degree of robustness and strong learning capability of the neural networks [Prabhakar and Henderson, 1992; Hwang and Henderson, 1992; Gu, 1997; Lankalapalli, *et al.*, 1997; Zulkifli and Meeran, 1999; Onwubolu, 1999; Ozturk and Ozturk, 2001]. A neural net can be thought of as being a reasoning black box, which inputs are the part information represented as vectors and sent to the net one-by-one, and which outputs are vectors that can indicate feature patterns, as illustrated in Fig. 2.15. Trained or self-adaptive neural networks, such as an MLFF perceptron or an ART net, are popularly used to recognise and classify features.

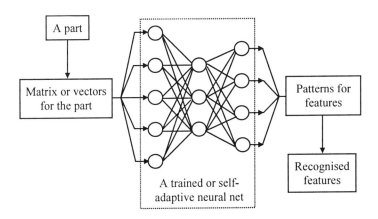

Fig. 2.15 The workflow for neural networks to recognise features.

Prabhakar and Henderson [1992] were the pioneers to apply the neural networks technique to the manufacturing feature recognition domain. In their approach, a part is first described as a graph. The graph is next encoded as a face adjacency matrix, and this matrix is input into a neural network for pattern recognition and classification. The dimension of the face adjacent matrix is $n \times n$, where n is the number of the faces in the part. Each row of the matrix, which includes n elements (faces), encodes the relevant information of a face. The diagonal element defines the properties for the face itself, and the other elements give the information of its relationships with other faces in the part. Gu [1997] developed a more detailed representation of the adjacency between faces, given as below (a few are illustrated in Fig. 2.16):

No adjacent relation	*0*
Planar face and planar face with 270° *convex*	*7*
Planar face and planar face with 90° *convex*	*4*
Planar face and surface with 90° *concave*	*1*
Planar face and surface with 270° *convex*	*2*
Planar face and planar face with non- 270° *convex*	*6*

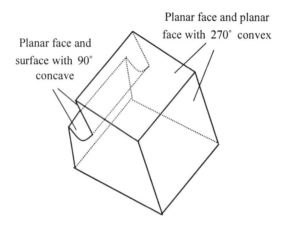

Fig. 2.16 The relationships between faces for neural networks-based feature recognition.

Onwubolu [1999] incorporated the geometry, loops and edges of faces to form face complexity codes, in which faces are classified as convex, concave or planar, edges as convex or concave, and loops as positive or negative. The convexity and concavity of faces and edges are similar to the concepts introduced in the rule- and graph-based approaches. A loop is defined as positive if it is visible when viewed from a main face, and negative if it is not visible from this viewpoint.

The neural networks-based approach is a promising approach to solve the interacting features and recognise new types of features. However, the linear or matrix input requirement of neural networks makes it difficult to input and represent all the required information in the graph representation of a part. The common standpoint for the above methods is to input the encoded information of a part as a whole into the neural networks. However, due to the diversity of the geometrical structures, the number of faces is different part-by-part. This makes it difficult to design a generic neural network to handle all parts, and it is quite difficult for some trained nets to be convergent. With the existence of complex interacting features in some parts, it is difficult to solely use neural networks to identify features. These drawbacks limit the practical applications of these methods. It is imperative to develop some preliminary steps to represent the information in a concise way for the input of the neural networks so as to achieve optimal efficiency in feature recognition.

To address this problem effectively, some neural networks-based hybrid methods have been developed [Nezis and Vosniakos, 1997; Wong and Lam, 2000]. Nezis and Vosniakos [1997] developed a two-stage method for feature recognition. In the first stage, based on the analysis of the convexity relationships between faces, ten heuristic rules were designed to separate a graph representation of a part into sub-graphs. The second stage employs a trained MLFF net to work on the sub-graphs for feature classification. Since the neural network handles the sub-graphs of features instead of the graph of a part, it is unified in structure and convenient in algorithm design. With the trained net, feature families with differences in geometric and topological structures and interacting features can be categorised effectively. In the work by Wong and Lam [2000], a two-stage neural network-based algorithm was designed for recognising orthogonal and non-orthogonal features. An orthogonal feature is defined to compose of a fixed number of faces and edges, and the adjacent faces are parallel or perpendicular to each other. A non-orthogonal feature does not have a fixed number of faces and edges, and the adjacent faces need not be parallel or perpendicular to each other. The first stage converts a non-orthogonal feature into an orthogonal feature. An MLFF is trained to facilitate this process. The second stage uses a volume decomposition method to generate multiple interpretations of the feature sets. In Table 2.2, some related works for the neural networks-based approach are summarised.

Table 2.2 A summary of some neural networks-based methods.

Works	Neural Networks	Major Characteristics	Recognised Features Types
Prabhakar and Henderson, 1992	MLPP	An adjacency matrix is used to code part topological and face geometrical information. Faces are categorised as plane faces and cylindrical faces, and edges are straight edges and circular edges.	Blind and open step/slot/pocket, groove, cut-out, bevel and keyway
Hwang and Henderson, 1992	MLPP	Each face has n neighbour faces, and a design part is converted to a 2D face set. According to the numbers of convex and concave edges in a face, a face score is computed using an evaluation formula. A face score graph is used as the input of networks for classification.	Blind and open step/slot/pocket
Gu, 1997	MLPP	More detailed topological information of faces and adjacent edges between faces are recorded to form an adjacency matrix for feature classification.	Blind and open step/slot/pocket, blind hole
Lankalapalli, *et al.*, 1997	ART 2	Face score vectors are computed based on the topology and geometry of their neighbour edges, inner loops and themselves. A feature type has a pattern of face score vector, which can be used for classification.	Tab, slot, protrusion, pocket, through and blind hole, boss, step and cross-slot
Nezis and Vosniakos, 1997	MLPP	Heuristic rules are employed to decompose a part as some sub-graphs to ease the recognition difficulty. Meanwhile, interacting relationships and variant features are handled to enhance the adaptability of the method.	Blind and open step/slot/pocket, and protrusion
Wong and Lam, 2000	MLPP	Two-stage neural networks-based algorithm is used for recognising orthogonal and non-orthogonal features.	Blind and open step/slot/pocket/hole, and boss

2.3 Volumetric Feature Recognition Scheme

The volumetric scheme, which can be classified as the convex hull approach and the volume growing/decomposition approach, recognises manufacturing features from a design part using volumetric operations.

2.3.1 *Convex hull approach*

The convex hull approach [Woo, 1977, 1982; Kim, 1990; Kim and Wilde, 1992; Kim and Wang, 2002; Owodunni and Hinduja, 2002] applies the convex hull operation and the regularised Boolean operations to decompose a solid model into volumetric features. The method proposed by Kim [1990] is described as follows.

- The convex hull $CH(P)$ of a polyhedron part P is the smallest convex set containing P.
- The regularised convex hull difference $CHD(P)$ of P is defined as $CHD(P) = CH(P) - P$.
- Conversely, P can be represented by the regularised difference between its convex hull ($CH(P)$) and its convex hull difference ($CHD(P)$), i.e., $P = CH(P) - CHD(P)$. The $CHD(P)$ of P is partitioned into maximal connected components. Let $D_i(P)$ be the ith regularised maximal connected component of $CHD(P)$, then $P = CH(P) - (\bigcup_i D_i(P))$.
- The above decomposition process is applied recursively to each $D_i(P)$. Finally, P is represented by a hierarchical tree, in which the non-terminal nodes are regularised difference and union operators and the terminal nodes are convex components.

Figure 2.17 is an example to illustrate the above process. $P = P^1 - (P^2 - (P^3 - P^4)) = ((P^1 - P^2) + P^3) - P^4$. That is, P consists of four features - P^1, P^2, P^3 and P^4. P^1 is a stock feature, P^2 and P^4 are subtractive features, and P^3 is an additive features. However, this method has a non-convergence problem for some parts (e.g., Fig. 2.18). In order to handle this problem, some remedial methods have been developed by Kim and Wilde [1992].

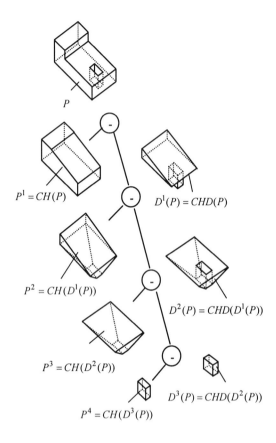

P

$P^1 = CH(P)$

$D^1(P) = CHD(P)$

$P^2 = CH(D^1(P))$

$D^2(P) = CHD(D^1(P))$

$P^3 = CH(D^2(P))$

$D^3(P) = CHD(D^2(P))$

$P^4 = CH(D^3(P))$

Fig. 2.17 A volumetric decomposition process based on the convex hull approach.

To recognise manufacturing features from cast-then-machined parts, a hybrid method based on a face pattern-based algorithm, a volume growing approach, and a convex hull approach, is presented by Kim and Wang [2002]. This method is unique in addressing parts made by multiple manufacturing processes, in which primary processes, such as casting, are used for realising the approximate shape of the parts, and secondary processes, such as machining operations, are used to generate the final shape of the parts. In this hybrid method, a face pattern-based algorithm is applied to identify simple features in a part, such as holes, pockets, slots, steps, etc., and a convex hull algorithm is applied to recognise complex manufacturing features in the part. The problem in this method is that features with uncommon shapes might be generated. Meanwhile, this approach cannot effectively support an user-defined feature taxonomy. These limitations obstruct the effective practical applications of this method but it brings a promising direction to handle parts with multiple manufacturing processes.

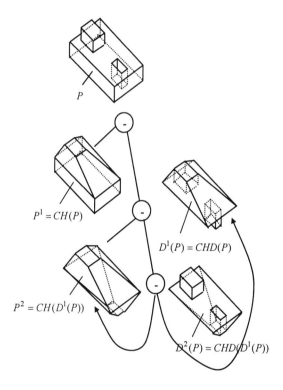

P

$P^1 = CH(P)$

$D^1(P) = CHD(P)$

$P^2 = CH(D^1(P))$

$D^2(P) = CHD(D^1(P))$

Fig. 2.18 A non-convergence problem for the convex hull approach.

2.3.2 *Volume growing/decomposition approach*

The volume growing approach [Wang and Chang, 1990; Chamberlain, *et al.*, 1993; Vandenbrande and Requicha, 1993; Xu and Hinduja, 1998] looks for hints of features, such as clusters of faces or loops from a part, and adds the volumetric features, which are converted from these hints through sweeping or volume enclosure operations, back to the part. The process continues until no hints and features can be detected in the incrementally grown part. In Vandenbrande and Requicha's approach [1993], the required and basic portions of the volumetric features (feature hints) are first determined in a delta_volume, which is the difference between a part and its original stock. The required portions are next extended along their TADs until they reach the boundary of the delta_volume. The extended portions are optional, and they bring forth the alternative plans (Fig. 2.19). Xu and Hinduja's method [1998] is based on the hint-based approach and the volume growing approach. Firstly, from the convex inner loops (feature hints), features can be detected. The corresponding volume forming features are next constructed. Finally, considering that the constructed volumes of the features may enclose materials that are present on the original stock, especially when a casting/forging process is involved, a volume verification process is executed through a set of Boolean operations on the stock and the feature volumes.

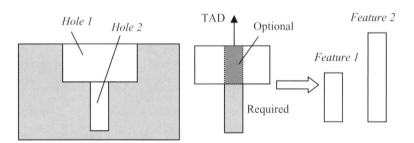

Fig. 2.19 Optional and required portions of a hole to generate alternative recognised features.

Based on delta_volumes, several volume decomposition approaches have been developed [Shah, *et al.*, 1994; Tseng and Joshi, 1994; Tseng and Lin, 1998; Sakurai, 1995, 1996; Dong and Vijayan, 1997a, b, c; Bezdek, *et al.*, 1999; Gaines and Hayes, 1999; Yui and Egbelu, 2000;

Kailash, *et al.*, 2001; Woo and Sakurai, 2002]. The approaches [Shah, *et al.*, 1994; Tseng and Joshi, 1994; Sakurai, 1995, 1996; Tseng and Lin, 1998; Dong and Vijayan, 1997a, b, c; Bezdek, *et al.*, 1999; Woo and Sakurai, 2002] decompose a delta_volume into volumetric units through face extension or half-space partition. These units are combined as varied sets of manufacturing features using some heuristic rules or algorithms (Fig.2.20).

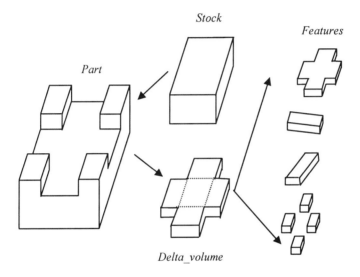

Fig. 2.20 Feature generation from a delta_volume using a volumetric decomposition approach.

Based on a minimisation of the total machining costs, and cutting tool and fixture utilisation costs, Dong and Vijayan [1997a] applied genetic algorithms to optimise the combination of the decomposed units to form features. The delta_volume is first decomposed as units. Some potential machining volumes are then formed through combining these units. Based on a minimisation of the cutting tools' working volumes (so as to reduce the total machining time and cost), different combinations of machining volumes are computed using genetic algorithms to find an optimal plan. This work is novel in introducing a combinatorial optimisation method to evaluate the manufacturing features and the relevant machining plans in quantity. However, this work assumed that the arrangement of the tools and fixture set-ups are fairly independent, which is not true in some cases. Woo and Sakurai's method [2002] can handle this deficiency. In their method, intermediate features called maximal features are first generated. By examining the possible set-up directions and machining sequences of the intermediate features, a set of features representing different interpretations can be obtained. A more comprehensive analysis of the relationships of the machining cost and the machining elements, such as tools, machines, set-ups, etc., to reflect the latest development for this problem is presented in Chapter 5.

For the application of a volume decomposition method to a complex part, much computing is involved to enumerate all the alternative interpretations. To address this problem, some methods utilise face information to guide and facilitate the volume decomposition/composition process. Yui and Egbelu's approach [2000] classifies the faces in each separated partition of a delta_volume as exposed, different hemispherical, and non-different hemispherical faces, as shown in Fig. 2.21. The exposed faces can be used directly to position a cutting tool to approach the stock. Different hemispherical faces can be approached by a tool starting from an exposed face. To machine a non-different hemispherical face, a newly exposed face, namely, a pseudo-exposed face, is added in the partition to split it further. Each partition is a manufacturing feature. By labelling a feature with one of its TADs, alternative interpretations are generated. Kailash, *et al.*, [2001] classified faces in a delta_volume as machining faces and non-machining faces. Depending upon the characteristics of the machining faces that have been detected through a comparison of the faces of the delta_volume with those of the part's stock, machining processes are associated. However, some desired alternative interpretations cannot be generated using these two methods.

In Gaines and Hayes's method [1999], the shapes of the cutting tools are considered during the feature recognition process. This work addresses a practical situation that the types of manufacturing features are adaptive in shop-floors. That is, with different cutting tools, the shapes of the manufacturing features to be extracted are different.

Therefore, the goal of this research is to propose a machining resource-adaptive recognition method based on the volumetric decomposition approach and a shape matching algorithm for a customised cutting tool. The developed system provides a tool editor for users to customise the desired shapes of the tool. With a defined tool, from the delta_volume of a part, algorithms have been developed to identify directions from which the tool can approach each face (TAD), and determine whether the cutting tool is capable of providing the required motions. To determine a TAD, the delta_volume is sliced into many 2D cross-sections, with which the cross-sectioned shape of the tool is compared. The tool is assigned for the part to produce the manufacturing feature if these two shapes match well. The difficulty of this method is that it cannot address the interacting features very well.

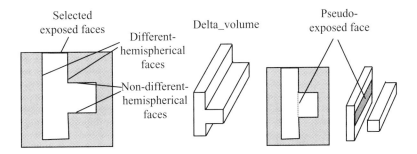

Fig. 2.21 Face classification for feature recognition.

2.4 Integration of Design-by-Feature and Feature Recognition

With the maturity of the design-by-feature systems, some feature recognition systems have been integrated with the design-by-feature systems to utilise the high-level design feature information in them. In some literatures, this approach is named as the "feature mapping technique" from design to manufacturing applications and domains.

Laakko and Mantyla's system [1993] is the first to achieve an incremental feature recognition approach from a design part by dynamically comparing the attributed graph of an evolved part with that of the previous part. During the design process of a part, the user can interactively modify either the solid model or the feature model of the part while the recognition system is kept updated with the changed one. An example to illustrate the operation of the incremental recognition process is shown in Fig. 2.22. However, the system does not provide a mechanism to decompose the complex structures of nodes caused by the interactions of design features into separated manufacturing features.

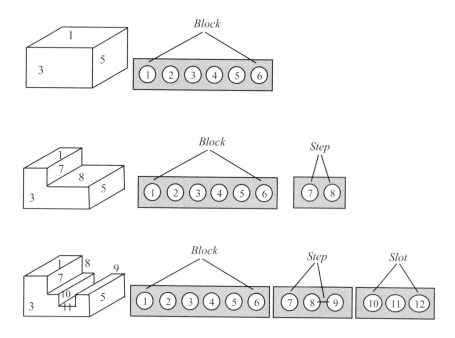

Fig. 2.22 An incremental integration of design-by-feature and feature recognition.

An intermediate geometric model described by De Martino, *et al.*, [1993, 1998] is used to integrate the design-by-feature and feature recognition systems. A feature-boundary processor of the system converts a design feature-based model into a geometric model, and a feature recogniser extracts manufacturing features from the geometric model using a graph-based approach. However, the high-level information of the design feature model has not been taken advantage of sufficiently, and the efficiency of the system is low due to the low-level geometric and topologic computing.

A heuristic reasoning algorithm developed by Han and Requicha [1997] describes a manufacturing feature recognition process based on a combination of the nominal geometry (feature hints), negative design features, tolerances and attributes of a design model. The feature recognition approach strives to produce a desirable interpretation of a part as quickly as possible. Alternative feature models can be generated on demand. In the recognition process, hint-based reasoning, blackboard architecture and uncertain reasoning techniques are combined. Negative design features provide strong evidences for manufacturing features. For instance, in Fig. 2.23, a negative design feature can be a manufacturing feature directly, or it has to be separated into several manufacturing features with the introduction of a positive feature. Negative design features need more considerations than other evidences, such as nominal geometry or attributes, because users tend to design with such features when they want them to appear in the final part, and therefore do not normally obliterate them after they have been introduced [Han and Requicha, 1997]. The advantage of this system is that it can utilise high-level design information to ease the recognition process. The shortcomings include the inability to recognise the machining volumes of additive design features, and the difficulty of determining the contribution of the strength of the evidence to the final decision.

A multiple feature conversion method to support a multi-disciplinary design environment was proposed by de Kraker, *et al.* [1995], and Bronsvoort and Noort [2004]. Engineers from different disciplines, such as product design, stress analysis, cost analysis, or process planning, can simultaneously work on the design of a product based on the multiple feature conversion method. In this method, a feature model is represented as two levels, namely feature and evaluated geometry. The product model is initially specified from a generic feature model. When a new view, for instance, cost analysis, is opened, feature conversion is performed using a central cellular model and view-specific information (Fig. 2.24). Modifications can be made in all views through changing the feature parameters, and adding or deleting feature instances in the views.

The constraints in all views can be maintained and propagated by the modelling system. However, the conversion method is based on solid modelling computation. Similar to the work by De Martino, *et al.*, [1993], the efficiency of the computation is not very high.

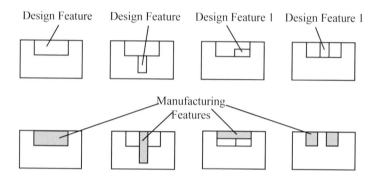

Fig. 2.23 Negative design features versus manufacturing features.

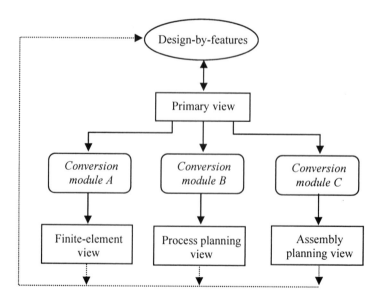

Fig. 2.24 Architecture of the multi-view feature system with primary and secondary views.

In the work reported by Suh and Ahluwalia [1995] and Perng and Chang [1997], modifications of interacting manufacturing features can be updated efficiently. A set of rules based on the geometric relationships between features has been devised to facilitate the propagation of the modifications. In their works, however, no methodology was developed to convert a design feature model into a manufacturing feature model, and the system cannot link design and manufacturing processes.

A geometric reasoning method proposed by Lee and Kim [1998, 1999] recognises manufacturing features from a design feature model based on STEP. Feature recognition is achieved through an incremental feature recognition approach, which not only keeps the design model consistent, but also incrementally extracts manufacturing features from design features as a design plan evolves. By combining the strengths of design-by-feature and feature recognition approaches, feature interactions and protrusion features can be efficiently handled. Through re-orientation, reduction and splitting operations, alternative interpretations are generated. However, in their work, certain types of interactions between features are not represented and handled, and complex manufacturing features, such as compound and replicate features cannot be recognised. Li, *et al.*, [2002] enhanced this method through summarising and manipulating complex feature relationships, and providing operations to generate multiple interpretations for features. The details of this method will be presented in Chapter 4.

Jha and Gurumoorthy [2000] reported an approach of automatic propagation of feature modification across domains, and multiple feature interpretation across domains. The approach is able to detect portions of a feature model that are not affected by the modifications made, which results in an efficient propagation of the modifications. The method can generate multiple feature interpretations for parts.

2.5 Summary

Table 2.3 gives a summary of some previous works. From the literature review, it can be observed that extensive attempts have been made in the area of manufacturing feature recognition, which is one of the essential modules to link product design and manufacturing for establishing a CE environment. However, the practical implementations of these methodologies are still far from satisfactory. This problem is considered as an imperative but tough research topic. To overcome the current bottleneck of this problem, breakthroughs need to be made, but not limited to, in the following three aspects.

Firstly, the capability of the methodologies for the recognition of interacting manufacturing features needs to be enhanced. A major cause that limits the current methodologies and systems is the complexity of the interacting features, which cause the recognition algorithms to be inaccurate. A single approach might be effective in certain sub-tasks in the interacting feature recognition process. Optimum efficiency and result, however, have not been achieved. Advanced intelligent and hybrid strategies that incorporate the advantages and diverse capabilities of various methods in the different phases of the recognition process is a promising research direction to address this problem.

Secondly, more efforts are needed to improve the adaptability of manufacturing feature recognition approaches. Features can be defined in various levels of abstractions and specifications, and different data exchange standards can be used to define different "schemas" for the features. The types of features to be chosen depend upon the feature taxonomy that a manufacturing cooperation would propose. The ability to generate alternative interpretations of the manufacturing features provides a potential way to overcome the shortcomings of the algorithms that are specific to the information of a part. Depending on the specific machining capabilities, conditions, and the data standard adopted, it is feasible for a manufacturing corporation to propose its manufacturing features definitions that are the most convenient and feasible. To support multiple workshop environments and applications, it is imperative to develop a methodology to provide multiple manufacturing interpretations for a design part to support a more generic application environment.

Thirdly, dynamic integration of design-by-feature and feature recognition approaches is necessary. It is important to establish an incremental feature recognition approach from an evolving design part that is constructed in a design-by-feature system, and to establish a dynamic and efficient re-recognition approach for a modified part through utilising the previous intermediate recognition information of the part, so as to support a dynamic integration of the two approaches.

Table 2.3 A summary of the developed feature recognition approaches.

Research Works	Methodologies										Results		Design Models			
	Depth Fillet	Decomposition	Volume Growing	Convex-hull	Genetic Algorithm	Uncertain Reasoning	Neural Net-based	Hint-based	Graph-based	Rule-based	Multiple Interpretations	Single Interpretation	Feature-based Design	Volume Rep	CSG	B-Rep
Woo, 1977				*								*			*	
Kyprianou, 1980										*		*				*
Choi, et al., 1984										*		*				*
Henderson, 1984										*		*				*
Lee and Fu, 1987												*			*	
Joshi and Chang, 1988									*			*				*
Corney and Clark, 1991									*			*				*
Gayanker and Henderson, 1990									*			*				*
Marefat and Kashyap, 1990									*			*				*
Sakurai and Gossard, 1990									*			*		*		*
Wang and Chang, 1990		*										*				*
Gadh and Prinz, 1992	*															

Table 2.3 A summary of the developed feature recognition approaches (cont'd).

Research Works	Design Models				Results		Methodologies									
	B-Rep	CSG	Volume Rep	Feature-based Design	Single Interpretation	Multiple Interpretations	Rule-based	Graph-based	Hint-based	Neural Network-based	Uncertain Reasoning	Genetic Algorithm	Convex-hull	Volume Growing	Decomposition	Depth Fillet
Hwang and Henderson, 1992	*				*					*						
Prabhakar and Henderson, 1992	*				*					*						
Chamberlain, *et al.*, 1993	*				*									*		
Fields and Anderson, 1993	*				*			*								
Laakko and Mantyla, 1993				*	*			*								
De Martino, *et al.*, 1993				*	*			*								
Vandenbrande and Requicha, 1993			*			*			*						*	
Shah, *et al.*, 1994			*			*									*	
Tseng and Joshi, 1994			*			*									*	
Gu, 1997	*				*					*						
Han and Requicha, 1997	*			*	*				*		*					
Regli, 1995; Regli, *et al.*, 1997	*					*		*	*							

Table 2.3 A summary of the developed feature recognition approaches (cont'd).

Research Works	Methodologies										Results		Design Models			
	Depth Fillet	Decomposition	Volume Growing	Convex-hull	Genetic Algorithm	Uncertain Reasoning	Neural Network-based	Hint-based	Graph-based	Rule-based	Multiple Interpretations	Single Interpretation	Feature-based Design	Volume Rep	CSG	B-Rep
Senthil kumar, *et al.*, 1996									*			*				*
Wu and Liu, 1996									*			*				*
Dong and Vijayan, 1997a, b, c		*			*						*			*		
Lankalpalli, *et al.*, 1997							*		*			*				*
Nezis and Vosniakos, 1997							*		*			*				*
Gao and Shah, 1998								*	*		*					
Han and Requicha, 1998(a)								*	*			*				*
Lee and Kim, 1998, 1999		*									*			*		
Wang and Kim, 1998				*								*		*		
Xu and Hinduja, 1998									*			*		*		
Zhang, *et al.*, 1998				*					*			*				*

Table 2.3 A summary of the developed feature recognition approaches (cont'd).

Research Works	Design Models				Results		Methodologies									
	B-Rep	CSG	Volume Rep	Feature-based Design	Single Interpretation	Multiple Interpretations	Rule-based	Graph-based	Hint-based	Neural Network-based	Uncertain Reasoning	Genetic Algorithm	Convex-hull	Volume Growing	Decomposition	Depth Fillet
Bezdek, *et al*, 1999			*		*										*	
Zulkifli and Meeran, 1999	*				*					*					*	
Gaines and Hayes, 1999			*		*											
Li, *et al.*, 2000	*				*			*	*	*						
Li, *et al.*, 2002						*									*	
Wong and Lam, 2000	*		*			*		*		*					*	
Yui and Egbelu, 2000					*										*	
Ozturk and Ozturk, 2001	*		*		*			*		*						
Owodunni and Hinduja, 2002			*		*								*			
Woo and Sakurai, 2002					*										*	
Li, *et al.*, 2003	*					*		*	*	*						
Bronsvoort and Noort, 2004				*	*		*									

Chapter 3

A Hybrid Method for Interacting Manufacturing Feature Recognition

Recognising interacting features from a design part is a major challenge in the feature recognition problem. It is difficult to solve this problem using a single reasoning approach or AI technique. In this chapter, a hybrid method based on feature hints, graph theory and a neural network - ART 2 net, that can recognise interacting manufacturing features, is presented. Case studies show that the developed hybrid method can achieve optimal efficiency and result by benefiting from the diverse capabilities of the three techniques in the different phases of the recognition process.

3.1 Introduction

Many works have been published in the area of manufacturing feature recognition. Among these approaches, the rule-based approach [Kyprianou, 1980; Henderson, 1984; Choi, *et al.*, 1984] uses pattern-matching techniques and expert systems to develop a set of feature recognition rules. The graph-based approach [Joshi and Chang, 1988; DeFloriani, 1989; Falcidieno and Giannini, 1989; Sakurai and Gossard, 1990; Gayanker and Henderson, 1990; Chuang, 1991; Corney and Clark, 1991; Fields and Anderson, 1993; Venuvinod and Wong, 1995; Sentil kumar, *et al.*, 1996; Tuttle, *et al.*, 1998; Ibrahim and McCormack, 2002] requires matching a graph of a part to a set of pre-defined feature sub-graphs using graph manipulation algorithms. These approaches can recognise isolated features easily but meet with difficulties when features overlap and interact in such a way that the geometry and topology of some entities in the features, such as faces, edges or vertices, have been destroyed or have disappeared completely. The hint-based approach [Vandenbrande and Requicha, 1993; Regli, 1995; Sommerville, *et al.*, 1995; Regli, *et al.*, 1997; Han and Requicha, 1998; Han, *et al.*, 1998] has been proposed to simulate the human's intuitive process of recognising interacting features. A hint is conceptualised to be a certain geometric pattern of an interacting feature left in the nominal geometrical model, even though the configuration of the feature might have been destroyed. The artificial neural networks-based approach [Prabhakar and Henderson, 1992; Hwang and Henderson, 1992; Lankalapalli, *et al.*, 1997; Nezis and Vosniakos, 1997; Zulkifli and Meeran, 1999; Onwubolu, 1999; Wong and Lam, 2000; Ozturk and Ozturk, 2001] has been proven to be very effective in recognising interacting features and new types of features due to its high degree of robustness and strong learning capability. However, each approach is effective in certain sub-tasks of the interacting feature recognition process, and optimum efficiency and result cannot be achieved through a single means.

A hybrid method based on feature hints, graph manipulations and a neural network has been developed to recognise interacting manufacturing features from a design part. Based on an Enhanced Attributed Adjacency Graph (EAAG) of a part, Face Loops (F-Loops), which are defined as generic feature hints, can be first extracted as clues of the interacting features of the part. The relationships between the F-Loops are next established according to the geometric and topological relationships between their entities. F-Loop Graphs (FLGs), which are the potential features, can then be built from the relationships between the F-Loops. Finally, these FLGs are input into an ART 2 neural network to be classified into different types of features. The definitions and manipulations of the F-Loops and the relationships between the F-Loops provide a unified solution to identify potential features (FLGs) from complex interacting and overlapping topological structures. An FLG is a simplified graph representation that is much easier for an ART 2 network to categorise into feature types, avoiding the high risk of failures when working on a complex graph of a part. By utilising the self-adaptive capability of an ART 2 neural network, this method can be adapted to recognise new types of interacting features. Therefore, by applying these techniques in the different sub-tasks of the feature recognition process, the recognition process for interacting features can be optimised. The flowchart of the hybrid method is presented in Fig. 3.1.

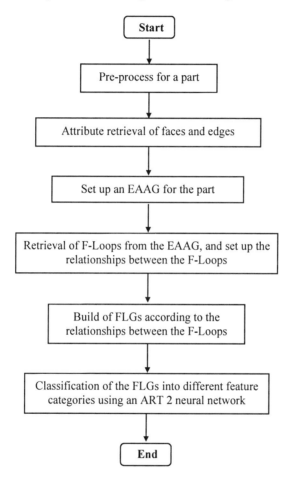

Fig. 3.1 The flowchart of the hybrid feature recognition method.

3.2 Enhanced Attributed Adjacency Graph

Since the geometric entities in the B-Rep model of a design part are explicitly represented so that it is convenient for recognition algorithms to act on, many existing feature recognition approaches are based on a B-Rep model. Among these approaches, some simplified graph schemes have been developed to extract useful geometric and topological information from the B-Rep model of a part to reduce the computation space. These schemes include the AAG [Joshi and Chang, 1988] and the EAAG, such as the symmetric boundary graph [DeFloriani, 1989], face-edge graph [Corney and Clark, 1991], cavity graph with virtual links [Marefat and Kashyap, 1990; Trika and Kashyap, 1994], etc.

An EAAG with extensions for some attributes of faces and edges has been implemented to facilitate the hybrid feature recognition method presented in this chapter. To set up the EAAG of a design part, a pre-process of two preliminary two steps is performed on the part.

3.2.1 Pre-process for generating EAAG

STEP 1: Handling of blending surfaces

A surface in a part is either to smoothly connect with its adjacent faces as a blending surface, or to sharply connect with its adjacent faces. In an EAAG, a blending surface is handled as a "smooth edge" if it is smoothly adjacent to two faces or a "smooth vertex" if it is smoothly adjacent to more than two faces. An example is illustrated in Fig. 3.2.

STEP 2: Handling of combinable planar faces or surfaces

Due to the existence of interacting features, a single face could be divided into several separated faces. These faces are *combinable* and can be restored as a single face in an EAAG if they satisfy the following two conditions simultaneously:

(1) If these faces are planar, they should share the same plane equation and have the same orientation; if these faces are surfaces, they can be joined to form a surface with the same equation; and

(2) These faces can be connected together to form a new face without being intersected by other faces in the part.

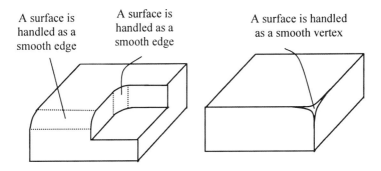

Fig. 3.2 A surface can be defined as a smooth edge or vertex.

To find the combinable machining faces, a "generate-test" method is employed:

(1) In the "generation" step, separate faces belonging to the same design feature in a part are identified to form a potential combinable face.

(2) In the "test" step, the potential combinable face is tested to determine whether it intersects other faces in the part. If it does not, these separated faces can be handled as a single face in an EAAG.

In Fig. 3.3, due to the interaction with *Slot 2*, two original faces of *Slot 1* are separated into $f_1 - f_2$ and $f_3 - f_4$. Since $f_1 - f_2$ and $f_3 - f_4$ can be combined without intruding any face in the part, they can be combined into one face. However, for $f_5 - f_6$ and $f_7 - f_8$, which are separated due to the interaction between features, they will interact with the island feature when they are combined into one face. Hence, they cannot be combined.

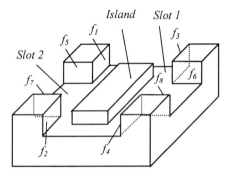

Fig. 3.3 A part with three design features.

3.2.2 *Establishment of EAAG*

After the pre-process, an EAAG of a design part, which can be represented as $EAAG = Graph(Face, Edge, Attribute)$, can be set-up through extracting the relevant information from the part. In the representation, *Face* represents a node in the graph, *Edge* represents the common edge between two adjacent faces, and *Attribute* represents some properties of a face and an edge that support the feature recognition approach. The faces and edges in the EAAG are classified in Table 3.1.

Table 3.1 Table 3 Classification of entities in an EAAG.

Entities	Classifications	
Faces	Machining face	Combinable machining faces
		Uncombinable machining faces
	Stock face	
Edges	Concave edge	
	Convex edge	Type I convex edge
		Type II convex edge
Attributes	Attributes for face	
	Attributes for edge	

The details in Table 3.1 are given as below and illustrated in Fig. 3.4.

- The faces of a part are partitioned into two classes: faces that are produced by the cutting tools (machining faces), and faces that coincide with the stock faces and they are untouched or partly removed by the cutting tools (stock faces) (Fig. 3.4 (a)-(c)).

- A convex edge in a part is classified as two types: a common convex edge between two machining faces (type I); a common convex edge either between a machining face and a stock face or between two stock faces (type II) (Fig. 3.4(d)).

- The attributes of a face include the properties of the face (planar face/surface, machining face/stock face), the boundary enclosure box of the face, the closeness of the face and the adjacent faces. The attributes of an edge include the connection type of the edge (sharp convexity, sharp concavity, smooth convexity, or smooth concavity (examples are shown in Fig.3.5)), the edge type (straight edge or curve), and a pair of vectors that are perpendicular to the edge and tangentially along the two adjacent faces of the edge respectively. This pair of vectors is later used to extract F-Loops from the EAAG (an example for a pair of vectors for an edge is shown in Fig. 3.4(e)).

3.3 Generation of Potential Features

Based on the established EAAG, F-Loops, which are defined as generic feature hints, can be extracted as clues for the interacting features of a part, and the relationships between the F-Loops are established according to the geometric and topological relationships between their geometric entities. FLGs, which are the potential features, can be built from the relationships between the F-Loops. These consecutive steps are described as follows.

3.3.1 *Identifications of F-Loops and their relationships*

Based on the EAAG space, generic feature hints, i.e., F-Loops, can be deduced. A valid F-Loop must satisfy the conditions defined in Table 3.2.

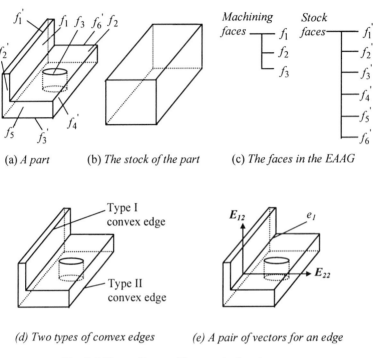

(a) *A part*   (b) *The stock of the part*   (c) *The faces in the EAAG*

(d) *Two types of convex edges*   (e) *A pair of vectors for an edge*

Fig. 3.4 The attributes of faces and edges in a part.

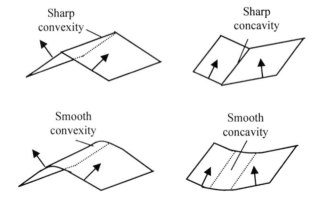

Fig. 3.5 Examples of four types of connections of two adjacent faces.

Table 3.2 Conditions for defining a valid F-Loop.

Cond-itions	Definitions & Illustrative Diagrams
(1)	An F-Loop is an undirected, linearly joined graph composing of a set of machining faces. Within an F-Loop, each machining face will not have more than two adjacent faces. In an F-Loop, the adjacent faces should have the same connection type, i.e., either concavity or convexity, and the *F-Loop* is defined as a concave or convex F-Loop correspondingly.
(2)	A set V of vectors can be established by crossing every pair of vectors in an F-Loop. The angles between each pair of adjacent vectors in V are calculated to form a set A. A valid F-Loop requires each element of A to be less than a given threshold θ, which is used to control the swing of the faces in the F-Loop. In this research, θ is assumed to be within $[0°, 5°]$.

Table 3.2 Conditions for defining a valid F-Loop (cont'd).

Cond-ition	Definitions & Illustrative Diagrams
(3)	Each face of an F-Loop can be extended along a common direction without intruding the interior solid volume of the design part. This common direction is the extended direction of the F-Loop. For an F-Loop, the boundary enclosure boxes (represented as rectangular planar faces) of the faces in the F-Loop projected perpendicularly to the extended direction should have a common intersection area. No common projected area for the three faces, so f_1-f_2-f_3 is not a valid F-Loop.
(4)	As a special case, a surface and its adjacent faces form an F-Loop. f_1, which is a surface, is a special F-Loop.

The F-Loops can be extracted from the EAAG of a design part using some manipulations. The procedure is described as follows:

(1) Each machining face in EAAG is associated with its neighbour stock faces and total neighbour faces according to following equation,

$$Rate(Face) = \frac{Number(Neighbour\_stock\ faces)}{Number(Neighbour\_faces)} \quad (3.1)$$

(2) The machining face with the highest value in the EAAG is chosen as a starting face to search for an F-Loop.

(3) The search process in the EAAG is continued according to Conditions (1)-(3) in Table 3.2, which are used to define and validate an F-Loop. During the search, if a face has more than one path that satisfies the definition of an F-Loop, the search process is continued according to the different paths to form several F-Loops, and the face is copied as elements for these F-Loops respectively.

(4) The above search process will continue, and any machining face that has been extracted as an element of an F-Loop will not be chosen as the starting face for a new F-Loop.

(5) After Steps (2) - (4) are completed, each surface, which satisfies Condition (4) in Table 3.2, and its adjacent faces in the EAAG are extracted to form an F-Loop.

(6) The relationships of the F-Loops are determined next.

In Fig. 3.6, the identification process of F-Loops is illustrated. In the EAAG of the part, each machining face is associated with a rate value. For instance, for f_1, it has four neighbour faces while two of them are stock faces. Therefore, the rate value for f_1 is 0.5. From f_1, an F-Loop, which consists of f_1- f_2- f_3, can be established. Similarly, another two F-Loops can be built as well. These F-Loops are validated according to the conditions defined in Table 3.2.

The relationships between two F-Loops that have adjacent edges or faces can be classified into six categories: simple adjacency, merged adjacency, cross, merged cross, nested adjacency and inseparable adjacency. The definitions of these relationships are given in Table 3.3. Fig. 3.7 shows a few examples to illustrate these relationships.

3.3.2 Identifications of FLGs

Based on the identified F-Loops and their relationships, F-Loop Graphs (FLGs) can be built. An FLG is a potential feature that can be further classified using a neural network.

F-Loops are manipulated according to the following steps to form FLGs, as illustrated in Fig. 3.8:

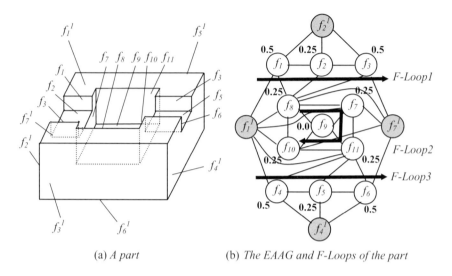

(a) *A part*   (b) *The EAAG and F-Loops of the part*

Fig. 3.6 A part and its EAAG and F-Loops.

Table 3.3 The relationships between two adjacent F-Loops.

Definitions	Definitions
Simple adjacency	If two concave F-Loops are convex adjacent, and their common edges do not form a closed edge loop, these two F-Loops are simple adjacent.
Merged adjacency	If two concave F-Loops have a common face while other conditions remain the same as in simple adjacency case, these two F-Loops are merged adjacent.
Cross adjacency	If a face in a concave F-Loop intersects with the interior area of a connected single face of another concave F-Loop, these two F-Loops are cross adjacent.
Merged cross adjacency	Merged cross adjacency is a special case of cross adjacency and it requires at least one face of each F-Loop to merge together as a single face and other requirements are kept the same as with cross adjacency.
Nested adjacency	When a concave (convex) F-Loop is defined as being nested in another F-Loop, all the convex (concave) adjacent edges form a closed edge loop.
Inseparable adjacency	If every face of a concave F-Loop is concave adjacent (or merged) with face(s) of another concave F-Loop, these two F-Loops are inseparable adjacent.

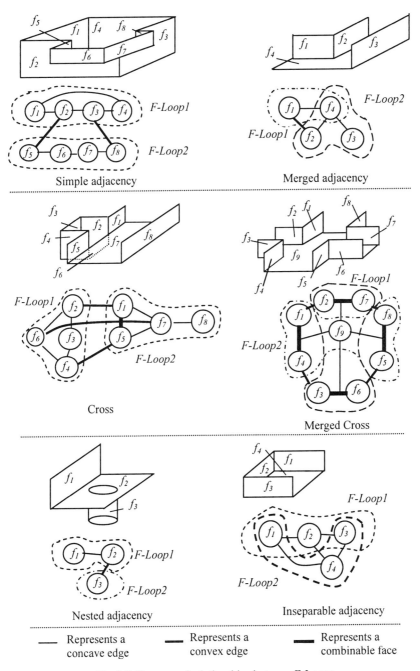

Fig. 3.7 Six types of relationships between F-Loops.

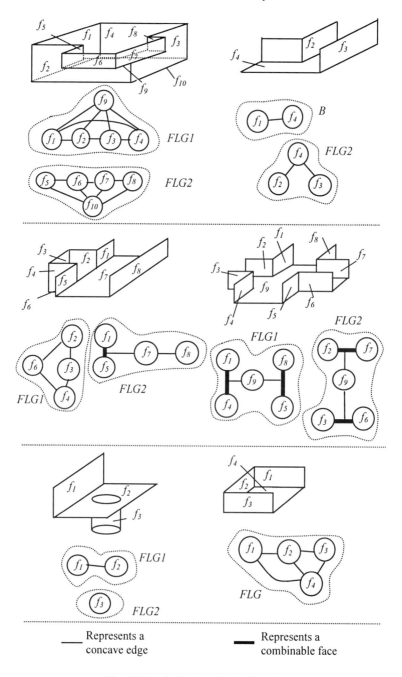

Fig. 3.8 Manipulations for forming FLGs.

(1) For the simple adjacency, the cross adjacency, and the nested adjacency relationships, two F-Loops are separated to form two FLGs.

(2) For the merged adjacency and the merged cross adjacency relationships, the merged faces are copied and allocated to each concave F-Loop, and the steps that follow are the same as the above Step (1).

(3) F-Loops with the inseparable adjacency are combined into a single FLG. If a portion of an F-Loop and another F-Loop have the inseparable adjacency, this part of the F-Loop is copied and combined with the second F-Loop to form a single FLG.

(4) In the EAAG of the part, if there is a face that is concave to all the faces in an F-Loop, they are combined as an FLG.

In an FLG, if there is a face that is concave to the other faces, this face is a base face. The FLGs can be classified into three types, as shown in Fig. 3.9:

Type 1:   In an FLG, there is a base face, and the other faces form an F-Loop.

Type 2:   In an FLG, there is a base face, and this face, together with the other faces, forms a single F-Loop in the FLG.

Type 3:   In an FLG, there is no base face, and all faces form an F-Loop.

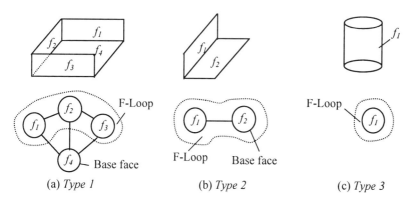

(a) *Type 1*   (b) *Type 2*   (c) *Type 3*

Fig. 3.9 Examples of three types of FLGs.

3.4 Neural Networks Classifier

A typical ART 2 net [Carpenter and Grossberg, 1987] with an architecture illustrated in Fig. 3.10 is used as a feature classifier. The ART 2 net involves three groups of neurons, namely, input processing units (F_1 layer), cluster units (F_2 layer) and reset units. The F_1 layer consists of six types of units, namely, w, x, u, v, p and q, and there are n numbers of each type of units (where n is the dimension of an input pattern). In the F_2 layer, the signals of the input units are combined and the similarities of the input vectors are compared with the weights of the cluster units. There are two sets of weights between the F_1 and F_2 layers. The bottom-up weights from F_1 to F_2 are denoted as b_{ij}, and the top-down weights from F_2 to F_1 are t_{kj}. Based on a "vigilance parameter" (ρ) in the reset units, the cluster units can decide whether to learn an input vector or not. In Fig. 3.10 and the algorithm to be presented next, a, b, c, d, e, θ, and α are the coefficients of the ART 2 net. The numbers of units in the F_1 and F_2 layers are n and m respectively. N_EP and N_IT are two learning rates.

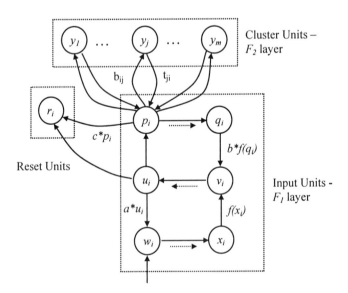

Fig. 3.10 A typical ART 2 architecture.

In the algorithm, $\| s_i \|$ and $f(x_i)$ are defined as follows:

$$\| s_i \| = \sqrt{s_1^2 + s_2^2 + ... + s_n^2} \qquad (3.2)$$

$$f(x_i) = \begin{cases} x_i & \text{if } x_i \geq \theta \\ 0 & \text{if } x_i < \theta \end{cases} \qquad (3.3)$$

The algorithm for the ART 2 net can be described as follows:

(1) Initialise a, b, c, d, e, θ, α and ρ (vigilance parameter), set the bottom-up weights b_{ij} and top-down weight t_{ji} as:

$$b_{ij} \leq \frac{1}{(1-d)\sqrt{n}} \qquad (3.4)$$

$$t_{ji} = 0 \qquad (3.5)$$

$$(\text{for } i = 1...n; j = 1...m).$$

(2) Perform Steps (3) to (13) for N_EP times.

   (3) For each input vector $x^k = \{x_1^k, x_2^k, ..., x_n^k\}$, perform Steps (4) to (12).

      (4) Update the units of F_1 layer:

$$u_i = 0 \qquad (3.6)$$
$$w_i = x_i^k \qquad (3.7)$$
$$p_i = 0 \qquad (3.8)$$
$$x_i = \frac{s_i}{e + \| s_i \|} \qquad (3.9)$$
$$q_i = 0 \qquad (3.10)$$
$$v_i = f(x_i) \qquad (3.11)$$

Update the units of the F_1 layer again:

$$u_i = \frac{v_i}{e+\|v_i\|} \tag{3.12}$$

$$w_i = s_i + au_i \tag{3.13}$$

$$p_i = u_i \tag{3.14}$$

$$x_i = \frac{w_i}{e+\|w_i\|} \tag{3.15}$$

$$q_i = \frac{p_i}{e+\|p_i\|} \tag{3.16}$$

$$v_i = f(x_i) + f(q_i) \tag{3.17}$$

(5) Compute signals to F_2 units:

$$y_j = \sum_{i=1}^{n} b_{ij} p_i \tag{3.18}$$

(6) While reset is true, perform Steps (7) to (8).

(7) Find the largest unit y_J in the F_2 layer.

(8) Check for reset:

$$u_i = \frac{v_i}{e+\|v_i\|} \tag{3.19}$$

$$p_i = u_i + dt_J \tag{3.20}$$

$$r_i = \frac{u_i + cp_i}{e+\|u_i\|+c\|p_i\|} \tag{3.21}$$

If $\|r_i\| < \rho - e$

$$y_J = -1 \ (inhibit \ J) \tag{3.22}$$

Reset is true, repeat (6).

If $\|r_i\| \le \rho - e$

$$w_i = s_i + au_i \tag{3.23}$$

$$x_i = \frac{w_i}{e+\|w_i\|} \tag{3.24}$$

$$q_i = \frac{p_i}{e+\|p_i\|} \tag{3.25}$$

$$v_i = f(x_i) + bf(q_i) \tag{3.26}$$

Reset is false, go to step (9).

(9) Perform Steps (10) to (11) for N_IT times.

(10) Update the weights for the winning unit J:

$$t_{Ji} = \alpha d u_i + \{1 + \alpha d(d-1)\}t_{Ji} \tag{3.27}$$

$$b_{iJ} = \alpha d u_i + \{1 + \alpha d(d-1)\}b_{iJ} \tag{3.28}$$

(11) Update the F_I activations:

$$u_i = \frac{v_i}{e+\|v_i\|} \tag{3.29}$$

$$w_i = s_i + a u_i \tag{3.30}$$

$$p_i = u_i + d t_{Ji} \tag{3.31}$$

$$x_i = \frac{w_i}{e+\|w_i\|} \tag{3.32}$$

$$q_i = \frac{p_i}{e+\|p_i\|} \tag{3.33}$$

$$v_i = f(x_i) + bf(q_i) \tag{3.34}$$

(12) Test stopping condition.

(13) Test stopping condition.

To employ the ART 2 net to classify the FLGs of an EAAG into different types of features, each FLG should be transformed into an input vector. Here, an FLG is organised as a 9-bit vector $x = \{x_i\}$, $i = 1\ldots 9$. The vector is determined as follows:

$$x_1 = \begin{cases} 3 & \text{The FLG is type 1} \\ -3 & \text{The FLG is type 2} \\ 0 & \text{The FLG is type 3} \end{cases}$$

$$x_2 = \begin{cases} 4 & \text{Existence of a base face} \\ -4 & \text{No existence of a base face} \end{cases}$$

$$x_3 = \begin{cases} 0 & \text{No existence of a base face} \\ 2 & \text{Base face is a polygonised boundary} \\ -2 & \text{Base face is not a polygonised boundary} \end{cases}$$

$$x_4 = \begin{cases} 1 & \text{F - Loop is convex and adjacent to stock faces} \\ -1 & \text{F - Loop is concave and adjacent to stock faces} \\ 0 & \text{F - Loop is self - closed} \end{cases}$$

$$x_5 = \begin{cases} 3 & \text{F - Loop is a surface F - Loop} \\ -3 & \text{Otherwise} \end{cases}$$

If $x_2 = 4$,

x_6 = Number of the adjacent edges between the faces in the F-Loop (except the base face for type 2) and the base face

$$x_7 = \begin{cases} 3 & \text{Only one common edge in } x_6 \\ -3 & \text{Several common adjacent edges} \\ 1 & \text{Several common non - adjacent edges} \end{cases}$$

else $(x_2 = -4)$, $x_6 = x_7 = 0$

x_8 = Total score of the edges in the F-Loop that are parallel to the extended direction of the F-Loop, $x_8 = \sum \text{score\_of\_edge}(i)$, where,

$$\text{Score\_of\_edge}(i) = \begin{cases} \text{Sharp convex edge} & 1 \\ \text{Sharp concave edge} & -1 \\ \text{Smooth convex edge} & 0.5 \\ \text{Smooth concave edge} & -0.5 \end{cases}$$

x_9 = Number of stock faces adjacent to an FLG

3.5 Computation Results

3.5.1 *Results for feature recognition*

Based on trials, the parameters of the ART 2 net are initialised as $a = 10$, $b = 10$, $c = 0.1$, $d = 0.8$, $e = 0$, $\theta = 0.1$, and $\alpha = 0.65$. N_IT is set as 1 for "slow learning" [Carpenter and Grossberg, 1987]. ρ, which is the "vigilance parameter", is used to control the similarity between the samples.

Eight basic types of features are illustrated in Fig. 3.11, and the FLG vectors for them are given in Table 3.4.

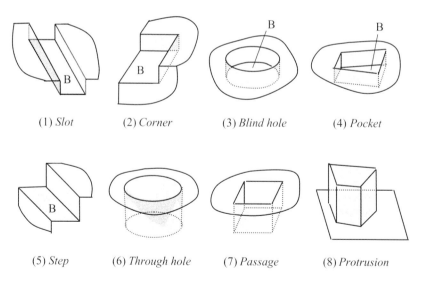

(1) *Slot*   (2) *Corner*   (3) *Blind hole*   (4) *Pocket*

(5) *Step*   (6) *Through hole*   (7) *Passage*   (8) *Protrusion*

Fig. 3.11 Eight basic types of features for classification.

Table 3.4 The FLG vectors for the eight basic types of features.

Features	x_1	x_2	x_3	x_4	x_5	x_6	x_7	x_8	x_9
Slot	-3	4	2	-1	-3	2	1	0	3
Corner	3	4	2	-1	-3	2	-3	1	3
Blind hole	3	4	-2	0	3	1	3	-0.5	1
Pocket	3	4	2	0	-3	4	-3	-4	1
Step	3	4	2	-1	-3	1	3	1	4
Through hole	0	-4	0	0	3	0	0	-0.5	2
Passage	0	-4	0	0	-3	0	0	-4	2
Protrusion	-3	4	2	0	-3	0	-3	4	1

Some results of the eight basic types of features were computed. Table 3.5 lists the results of the different epochs (*N_EP*) with the same ρ (ρ = 0.995). In Table 3.5, the number of each type of feature is randomly generated by the ART 2 algorithm. Features with the same number belong to the same feature family. When *N_EP* = 1, the algorithm was carried out for one epoch and eight features were categorised into five families. When *N_EP* = 45, the eight features were distinguished into eight basic types and the result from the algorithm was stable.

Table 3.5 Computation results for eight basic types of features (ρ = 0.995).

Features	*N_EP* = 1	*N_EP* = 15	*N_EP* = 30	*N_EP* = 45	*N_EP* = 60
Slot	1	6	6	8	8
Corner	1	1	1	1	1
Blind hole	2	2	2	2	2
Pocket	1	7	7	7	7
Step	2	6	6	6	6
Through hole	3	3	3	3	3
Passage	4	4	4	4	4
Protrusion	5	5	5	5	5

Twenty-one features (including the eight basic features), as shown in Fig. 3.12, are input into the ART 2 net for classification. The computation results are shown in Table 3.6.

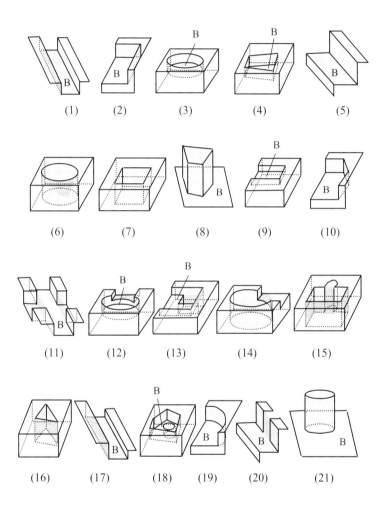

Fig. 3.12 Twenty-one features to be classified.

Table 3.6 The computation results for the twenty-one features.

Feature	$\rho = 0.988 - 0.992$	$\rho = 0.993 - 0.996$
Slot	1, 11	1, 11, 17
Corner	2, 4, 9, 10, 13, 17, 18	2, 9, 10, 13, 19
Blind hole	3, 12, 19, 21	3, 12
Pocket	2, 4, 9, 10, 13, 17, 18 (same as corner)	4, 18
Step	5, 20,	5, 20
Through hole	6, 14	6, 14
Passage	7, 15, 16	7, 15, 16
Protrusion	8	8, 21

Table 3.7 The recognition results of two practical parts.

Parts	F-Loops		Surface F-Loops		FLGs
	Concave	Convex	Concave	Convex	
1	16	7	6	0	23
2	6	2	7	0	13

Parts	Features							
	1	**2**	**3**	**4**	**5**	**6**	**7**	**8**
1	5	2	4	1	1	4	1	5
2	2	0	0	0	4	5	0	2

Features **1** – slot, **2** – corner, **3** – blind hole, **4** – pocket,
     5 – step, **6** – through hole, **7** – passage, **8** – protrusion

Fig. 3.13 shows more cases of features recognised using this approach. Two practical parts shown in Fig. 3.14 and Fig. 3.15 were recognised using the algorithm and the results are listed in Table 3.7.

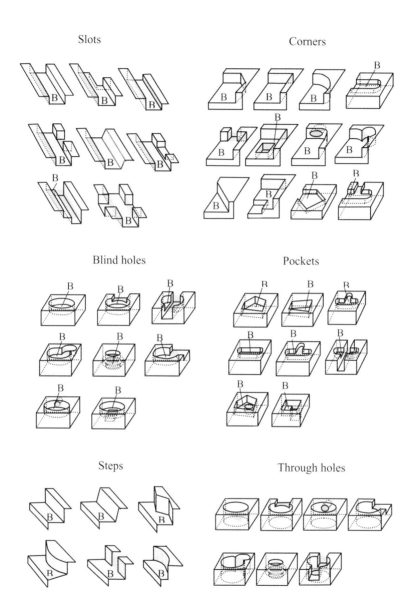

Fig. 3.13 More features that are recognised using this approach.

Passages   Protrusions

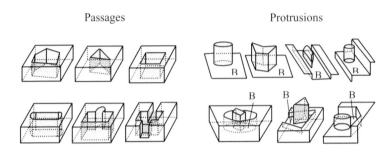

Fig. 3.13 More features that are recognised using this approach (cont'd).

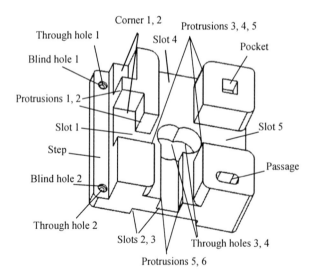

Fig. 3.14 The first part with recognised features.

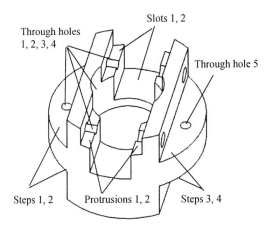

Fig. 3.15 The second part with recognised features.

Table 3.8 Comparisons between ART 2 and MLFF nets for feature recognition.

Neural Networks	Iteration time[*]	Number of feature categories	Convergence
ART 2	< 1s	8	Easy (1 trial)
MLFF	14s	8 (error[**] = 0.03584)	Difficult (>20 trials)

[*] Tested using an Intel Pentium II 350 micro-computer environment.

[**] The definition of the error is the Root Mean Square (RMS) between the trained and practical outputs.

3.5.2 *Result comparisons*

(1) Comparison with MLFF

A MLFF net with a back propagation training algorithm has been reported to be used for feature recognition [Gu *et al.*, 1995; Nezis and

Vosniakos, 1997]. A MLFF net was initially attempted in this research for the classification of features. The comparison results between the ART 2 net and the MLFF net for the eight basic types of features are shown in Table 3.8.

(2) Comparisons with other approaches for recognising interacting features

This approach was compared with several other approaches reported in the literature from the perspective of recognising interacting features and adaptability for recognising new types of interacting features. Since the executive codes of the approaches from the literature and being compared here are not available for testing on the features in the case studies, the analysis of the adaptability for recognising new types of interacting features for these approaches is given qualitatively. The feature interacting relationships for the comparisons are taken from [Gao and Shah, 1998] and shown in Fig. 3.16. Table 3.9 shows the comparisons.

3.6 Summary

In this chapter, a hybrid method based on feature hints, graph manipulations and an artificial neural network has been presented to recognise interacting manufacturing features from a design part. Based on the EAAG of a part, F-Loops, which are defined as generic feature hints, can be first extracted as clues for interacting features of the part. The relationships between the F-Loops are next established according to the adjacent relationships between their geometric entities. FLGs, which are the potential features, can then be built from the relationships between the F-Loops. Finally, the FLGs are input into an ART 2 neural network to be classified into different types of features.

The main characteristics of the method include the following two aspects:

Through generalising the concept of feature hints as F-Loops and summarising the interacting relationships between F-Loops, a systematic manipulation process has been developed to deduce potential features (FLGs). The definitions and manipulations of F-Loops and the relationships between F-Loops provide a unified solution to identify potential features from complex interacting and overlapping topological structures. An FLG is a kind of simplified graph representation that is much easier for the ART 2 network to handle and this avoids the high risk of failures for a complex graph of a part. Therefore, this methodology can facilitate the recognition of complex interacting manufacturing features.

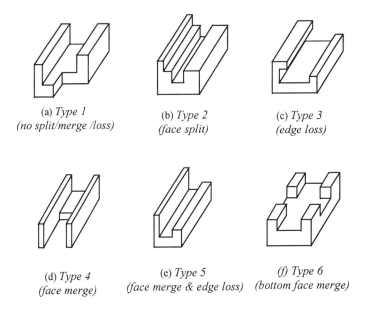

(a) *Type 1*
(no split/merge /loss)

(b) *Type 2*
(face split)

(c) *Type 3*
(edge loss)

(d) *Type 4*
(face merge)

(e) *Type 5*
(face merge & edge loss)

(f) *Type 6*
(bottom face merge)

Fig. 3.16 Feature interactions classification.

Neural networks, which are applicable in the pattern recognition domain, are suited for feature classification, self-adaptable to new types of interacting features, but not efficient in handling complex input patterns. This approach integrates the ART 2 network as the feature classifier with the inputs of simplified representations of potential features to enhance the performance of the employed network. By utilising the self-adaptive capability of the ART 2 neural network, this method can be adapted to recognise new types of interacting features. Meanwhile, by utilising the characteristics of several techniques in the different sub-tasks of the feature recognition process, the feature recognition process can achieve optimal efficiency and result.

Table 3.9 A comparison of different approaches for recognising interacting features.

Research	Capability for Recognising Interacting Features						Adaptability for New Features
	Type1	Type2	Type3	Type4	Type5	Type6	
Joshi and Chang, 1988	*						Weak
Marefat and Kashyap, 1990	*	*			*	*	Weak
Corney and Clark, 1991	*						Weak
Tseng and Joshi, 1994; Tseng and Lin, 1998	*	*	*	*	*	*	Weak
Senthil kumar, *et al.*, 1996	*	*		*		*	Weak
Dong and Vijayan, 1997a, b,_c	*	*	*	*	*	*	Weak
Nezis and Vosniakos, 1997	*		*	*			Strong
Lankalapalli, *et al.*, 1997	*		*			*	Medium
Gao and Shah, 1998	*	*	*	*	*	*	Weak
Han and Requicha, 1998	*			*			Weak
Zhang, *et al.*, 1998	*	*		*		*	Weak
This approach	*	*	*	*	*	*	Strong

Chapter 4

Integration of Design-by-Feature and Manufacturing Feature Recognition

Design is an interactive, dynamic and recurring process. Achieving reasonable levels of interactions between design and downstream manufacturing processes can help a designer to attempt alternative plans and evaluate manufacturability and machining cost during the initial design phase. For a CAD system that is not dynamically integrated with CAPP and CAM systems, a modification of a design part with manufacturing flaws would require a total re-work of the manufacturing interpretation of the part. This increases the design cost and the overall lead-time from design to manufacturing. Hence, it is essential to develop a dynamic integration strategy to link design-by-feature and manufacturing feature recognition approaches.

In this chapter, an approach is presented to recognise manufacturing features from a design model, which is created using a design-by-feature system. Based on the definition and analysis of feature-to-feature interacting relationships, a feature recognition processor can translate a design feature tree of a part into a manufacturing feature tree, in which a single interpretation is stored. A re-recognition process can efficiently update the manufacturing feature tree after adding, deleting, or modifying design features. Based on geometric operations, alternative manufacturing feature interpretations from a manufacturing feature tree are generated.

4.1 Introduction

Feature-based modelling can facilitate design and manufacturing applications through associating shape macros with semantic information for these activities. Features are domain-dependent. Design features created in a design-by-feature system contribute to the design construct basis of a part in terms of additive or subtractive volumes, and their semantics reflect the design intent and function. On the other hand, manufacturing features are normally formed from a manufacturing stock in terms of subtractive volumes, and they are logically connected with manufacturability analysis and process planning activities, such as fixture planning, machines and cutting tools selection, and machining operations planning. Manufacturing feature recognition from a design model can provide a seamless interface between design and manufacturing applications. Early systems, though able to recognise manufacturing features from solid models, are burdened with insufficient design information and low-level geometric computing.

With the popularity of the design-by-feature systems in product design, methods for converting a design feature model to a manufacturing feature model have been developed recently to enhance the recognition process through utilising the richer information of the design feature models [Laakko and Mantyla, 1993; Martino, *et al.*, 1993, 1998; de Kraker, *et al.*, 1995; Suh and Ahluwalia, 1995; Han and Requicha, 1997; Perng and Chang, 1997; Jha and Gurumoorthy, 2000; Bronsvoort and Noort, 2004]. Based on design features and their auxiliary information, manufacturing features can be translated from design features efficiently. The design process information reflected in the design-by-feature model of a part, which is incremental through introducing new design features in it, can be utilised to interpret the part in terms of manufacturing features during the design evolution process. This would allow a designer to attempt alternative design plans and evaluate the manufacturability dynamically. This technique can be enhanced in the following aspects:

- Representing and recognising interacting features is still a bottleneck that hinders this technique. An elaborate strategy is imperative for recognising features with complex interacting relationships.

- When a design part with manufacturing flaws is modified, the information of the previous intermediate recognition process could be utilised to facilitate the updating process of the manufacturing feature model, and a total re-work can be avoided. However, few researchers have made contributions towards this direction.

- Most of the feature recognition approaches interpret a design part as a single set of manufacturing features. However, with the interactions between features and the variations of manufacturing applications, a design part can be represented as more than one interpretation of manufacturing features. The ease and feasibility to machine a design part in a specific manufacturing condition depends on the interpretation of the manufacturing features that has been chosen. To support a generic application environment, it is imperative for a recognition algorithm to generate the alternative interpretations from a design part.

This chapter addresses these issues and presents an approach to recognise manufacturing features from a design model, which has been created using a design-by-feature system. Feature-to-feature interacting relationships in a volume representation are categorised. Based on these relationships, a feature recognition processor traverses the design feature tree of a part and translates the design feature tree into a manufacturing feature tree, in which a single interpretation is stored. A re-recognition process can efficiently update the manufacturing feature model of a part after any addition, deletion or modifications of its design features. Depending on the properties of the manufacturing feature tree, through geometric operations, alternative interpretations for the part are generated. The manufacturing feature is extended to represent and store these interpretations.

4.2 Features and Their Relationships

4.2.1 *Feature models*

In a design-by-feature system, the design feature tree of a part is a constructive tree used to organise design features of the part in a hierarchical structure according to its evolving process. The root of the tree is the construct design basis of the part. Each intermediate node in the tree is a Boolean union or difference operator, and its corresponding leaf is an additive design feature volumetrically added onto the part, or a subtractive design feature volumetrically removed from the part. An auxiliary design feature is attached to the root or a leaf as its auxiliary part. Each feature is associated with one or more datum planes or axes, surface tolerances, attributes, dimensions, etc. In this chapter, the shape of a feature is constrained as a swept volume through sweeping a 2D sketch along an "extended direction". The faces of a design feature that originate from the 2D sketch and are perpendicular to the extended direction are denoted as the "bases". The other faces in the feature, which are defined as analytic surfaces, i.e., planar, conical, spherical and toroidal, are the "silhouette faces". Fig. 4.1 shows these concepts.

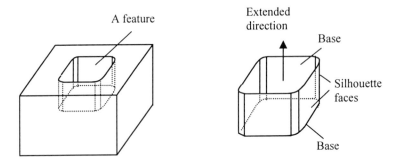

Fig. 4.1 A feature and some of its geometric elements.

STEP AP 224 (ISO 10303-224, 1996) specifies the manufacturing information and process plans using manufacturing features to machine discrete mechanical parts. The features defined in STEP AP 224 are shown in Fig. 4.2.

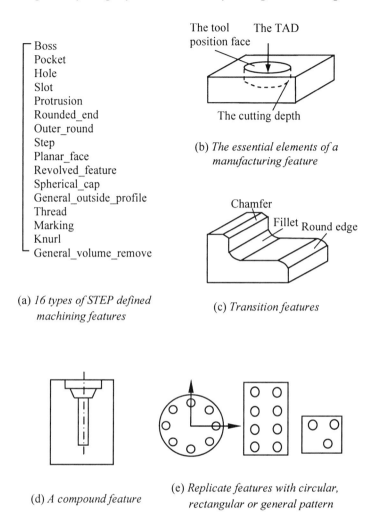

(a) *16 types of STEP defined machining features*

Boss
Pocket
Hole
Slot
Protrusion
Rounded_end
Outer_round
Step
Planar_face
Revolved_feature
Spherical_cap
General_outside_profile
Thread
Marking
Knurl
General_volume_remove

(b) *The essential elements of a manufacturing feature*

(c) *Transition features*

(d) *A compound feature*

(e) *Replicate features with circular, rectangular or general pattern*

Fig. 4.2 STEP AP 224 defined manufacturing features.

The geometry of a design feature without interactions with other features can be directly mapped to that of a manufacturing feature. However, a feature recognition processor stills need to consider the following observations:

- In some cases, essential machining elements of a manufacturing feature may not be obtainable. For example, a feature created along an extended direction might have several TADs (Tool Approach Directions), and these TADs should be determined by the feature recognition processor.

- Positioning a design feature in a part might cause the initial geometry of the feature to be changed after the Boolean operations. The interactions between the design features might also cause geometric variations from the initial ones.

- A design feature tree reflects a constructive process, and it may contain a few additive design features. A manufacturing feature model, on the other hand, reflects a decomposition process, where the manufacturing features are subtractive in volume. In AP 224, additive features including protrusive features and bosses are defined. However, their corresponding machining volumes are not expressed explicitly. To machine these features, the recognition processor should recognise the machining volumes of additive features as manufacturing features.

- Considering the varied manufacturing processes in different applications, a design feature might be converted into several reasonable and practical manufacturing features. For instance, a compound design feature might be mapped as either a compound manufacturing feature or several single features.

A few examples in Fig. 4.3 and Fig. 4.4 illustrate some of the above cases. In Fig. 4.3(a), a design pocket becomes a manufacturing step after it joins the part, and therefore, most of the design and manufacturing parameters of the feature are different. In Fig. 4.3(b), due to the interaction between design features, a new manufacturing feature is generated. In Fig. 4.3(c), in order to manufacture an additive design feature, the surrounding volume of the feature will need to be computed to form a manufacturing feature. In Fig. 4.4(a), either the slot or the hole can be machined first, and therefore the alternative machining sequences for this part bring forth two sets of manufacturing features with different hierarchies and volumes. In Fig. 4.4(b), the slot has two cutting TADs, so that an alternative set-up and process plan can be generated. In Fig. 4.4(c), considering the varied manufacturing conditions, a compound feature can be separately machined as several simple features.

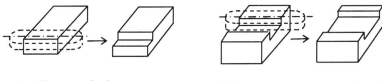

(a) *A design pocket becomes a
manufacturing step in a part*

(b) *Two design slots are combined
as a manufacturing slot*

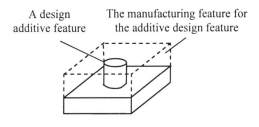

(c) *The manufacturing feature of
an additive design feature*

Fig. 4.3 In some cases, manufacturing features are different from their
corresponding design features.

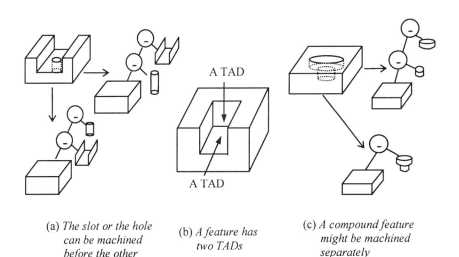

(a) *The slot or the hole
can be machined
before the other*

(b) *A feature has
two TADs*

(c) *A compound feature
might be machined
separately*

Fig. 4.4 A design part can be represented as several alternative interpretations.

4.2.2 *Interacting relationships between features*

The interacting relationships between a volumetric feature FE_2 added to a part P (P' is the updated part after the addition of FE_2) and a volumetric feature FE_1 in P are defined in Tables 4.1 and 4.2, and some examples are shown in Fig. 4.5. The symbols are defined as follows:

- ∂ represents the boundary set of a volumetric feature or partition. ϕ represents a void set. $\subset$ and $\not\subset$ represent belonging to and not belonging to a subset respectively.

- $\cap$, $\cup$, $-$ represent the Boolean intersection, union, and difference operators respectively.

- In the adjacency, splitting, disconnection, interference and penetration relationships, FE_1 is divided into k_1 volumetric partitions by FE_2, and FE_2 is divided into k_2 volumetric partitions by FE_1. These partitions belong to one of the following three sets: $R = \{r_{i(1 \leq i \leq L)} \mid r_i \in (FE_1 - FE_2)\}$, $S = \{s_{i(1 \leq i \leq M)} \mid s_i \in (FE_1 \cap FE_2)\}$ and $T = \{t_{i(1 \leq i \leq N)} \mid t_i \in (FE_2 - FE_1)\}$, where L, M, and N represent the number of partitions in R, S and T respectively. $L + M = k_1$ and $M + N = k_2$. For two adjacent partitions, from R and S, denoted as r_j and s_l, p is the number of the adjacent faces between r_j and s_l. For two adjacent partitions from S and T, denoted as s_j and t_l, q is the number of adjacent faces between s_j and t_l.

4.3 Manufacturing Features Recognition Processor

The feature recognition processor analyses the interacting relationships between each design feature in a design feature tree with the manufacturing features in an incrementally evolved intermediate manufacturing feature tree. The flowchart of the algorithm is shown in Fig. 4.6, where the manufacturing feature being traversed[*] is denoted as

---

[*] The traverse operation follows the definition for a tree data structure. To traverse a feature tree means each feature in the tree is visited once and only once.

MF , the design feature that is being traversed is DF , the stock of a MF tree is the manufacturing stock S , a design feature tree is a DF tree, and a manufacturing feature tree is an MF tree. In a MF tree, S is additive and the other features are subtractive in volume.

Table 4.1 The interacting relationships between features.

Relations	Conditions	Descriptions
Abutment	$FE_2 \cap FE_1 = \phi$, $\partial(FE_2) \cap \partial(FE_1) \neq \phi$.	FE_2 and FE_1 are abutted.
Adjacency	$FE_2 \cap FE_1 \neq \phi$, $FE_2 \not\subset FE_1$, $FE_1 \not\subset FE_2$, $N = 1$, $t_1 \cap P \neq \phi$, t_1 and each s_i satisfies $q > 1$.	FE_2 is adjacent.
	$FE_2 \cap FE_1 \neq \phi$, $FE_2 \not\subset FE_1$, $FE_1 \not\subset FE_2$, $L = 1$, η and each s_i satisfies $p > 1$.	FE_1 is adjacent.
Splitting	$FE_2 \cap FE_1 \neq \phi$, $FE_1 \not\subset FE_2$, $FE_2 \not\subset FE_1$, $N = 1$, $t_1 \cap P \neq \phi$, t_1 and each s_i satisfies $q = 1$.	FE_2 is split.
	$FE_2 \cap FE_1 \neq \phi$, $FE_2 \not\subset FE_1$, $FE_1 \not\subset FE_2$, $L = 1$, η and each s_i satisfies $p = 1$.	FE_1 is split.
Disconnect-ion	$FE_2 \cap FE_1 \neq \phi$, $FE_2 \not\subset FE_1$, $FE_1 \not\subset FE_2$, $N > 1$, for each $t_i : t_i \cap P \neq \phi$.	FE_2 is disconnected.
	$FE_2 \cap FE_1 \neq \phi$, $FE_2 \not\subset FE_1$, $FE_1 \not\subset FE_2$, $L > 1$.	FE_1 is disconnected.
Replicate Pattern	FE_1 is copied as several features $FE_{2..n}$, and these features are constrained to satisfy certain arrangement patterns.	$FE_{1..n}$ form a replicated feature ($FE_{1..n}$, *pattern*)

FE_1 and FE_2 are holes

Co-axis	The central axes of FE_1 and FE_2 are in the same line.	FE_1 and FE_2 are co-axial.

Table 4.1 The interacting relationships between features (cont'd).

Relations	Conditions	Descriptions
FE_1 and FE_2 are slots		
Co-symmetry	The symmetrical datum planes of FE_1 and FE_2 are in the same plane.	FE_1 and FE_2 are co-symmetrical.
Both of FE_1 and FE_2 are either holes or slots		
Virtual linking	The faces of FE_1 and FE_2 can be extended and linked as a single face without intruding the other faces in the part.	FE_1 and FE_2 can be virtually linked.
For the following nesting, interference and penetration, FE_2 is additive and FE_1 is subtractive, or FE_2 is subtractive and FE_1 is additive		
Nesting	$FE_1 \subset FE_2$.	FE_1 nests on FE_2 .
	$FE_2 \subset FE_1$.	FE_2 nests on FE_1 .
Interference	$FE_2 \cap FE_1 \neq \phi$, $FE_2 \not\subset FE_1$, $FE_1 \not\subset FE_2$, $s_i \cap P' \neq \phi$ for each $s_i (s_i \in S)$ (FE_2 is additive, and FE_1 is subtractive).	FE_2 interferes with FE_1 .
Penetration	$FE_1 \cap FE_2 \neq \phi$, $FE_2 \not\subset FE_1$, $FE_1 \not\subset FE_2$, $s_i \cap P' = \phi$ for each $s_i (s_i \in S)$ (FE_2 is subtractive, and FE_1 is additive).	FE_2 penetrates FE_1 .
FE_1 and FE_2 are additive		
Overlapping	$FE_2 \cap FE_1 \neq \phi$, $FE_2 \not\subset FE_1$, $FE_1 \not\subset FE_2$.	FE_1 and FE_2 overlap.
	$FE_2 \cap FE_1 = \phi$ and $\partial(FE_2) \cap \partial(FE_1) \neq \phi$.	

Table 4.2 The mutual interacting relationships between FE_1 and FE_2 in the adjacency, splitting and disconnection categories.

Relationships	FE_1 is adjacent	FE_1 is split	FE_1 is disconnected
FE_2 is adjacent	√	√	√
FE_2 is split	√	√	√
FE_2 is disconnected	√	√	√

√ There are cases between the items of the corresponding row and column.

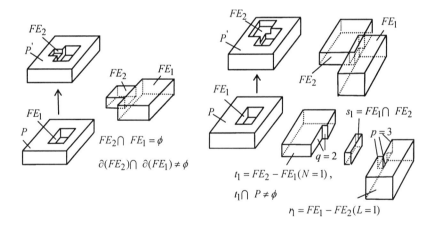

(a) FE_2 and FE_1 are abutted

(b) FE_2 is adjacent, and FE_1 is adjacent

Fig. 4.5 The interacting relationships between features.

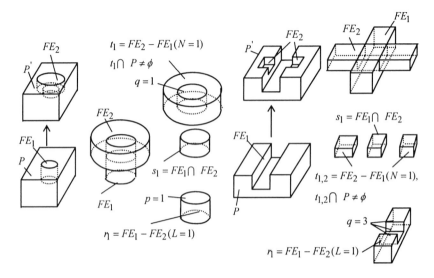

(c) FE_2 is adjacent, and FE_1 is split   (d) FE_2 is disconnected, and FE_1 is adjacent

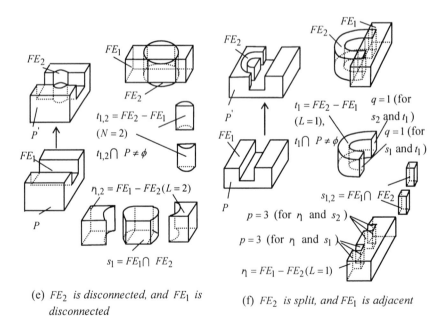

(e) FE_2 is disconnected, and FE_1 is
    disconnected

(f) FE_2 is split, and FE_1 is adjacent

Fig. 4.5 The interacting relationships between features (cont'd).

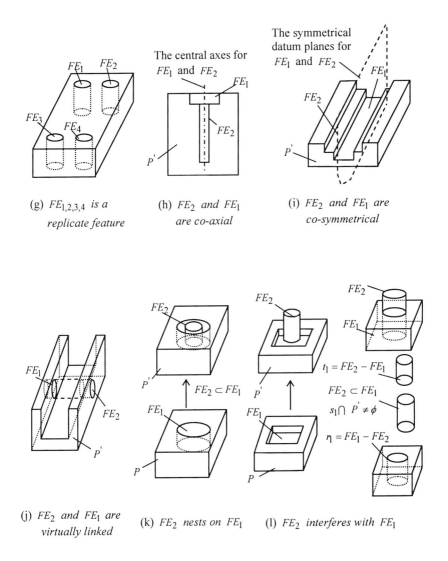

(g) $FE_{1,2,3,4}$ *is a replicate feature*

(h) FE_2 *and* FE_1 *are co-axial*

(i) FE_2 *and* FE_1 *are co-symmetrical*

(j) FE_2 *and* FE_1 *are virtually linked*

(k) FE_2 *nests on* FE_1

(l) FE_2 *interferes with* FE_1

Fig. 4.5 The interacting relationships between features (cont'd).

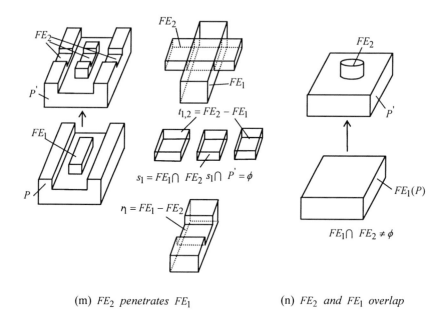

(m) *FE$_2$ penetrates FE$_1$*   (n) *FE$_2$ and FE$_1$ overlap*

Fig. 4.5 The interacting relationships between features (cont'd).

In the algorithm, there are mainly two nested traversal operations to translate a *DF* tree to a *MF* tree, followed by a re-recognition of a modification process of the design part (if any modification is imposed). An "outer traversal operation" traverses the *DF* tree backwards, starting from its root, and this process is only carried out once. An "inner traversal operation" traverses the *MF* tree backwards, starting from its root. For every *DF* traversed in the outer traversal operation, the whole cycle of the inner traversal operation will be performed once. After analysing the interacting relationships between *DF* and every *MF* traversed in the inner traversal operation, *DF* is handled and the *MF* tree is updated. The main steps highlighted in Fig. 4.7 are described next.

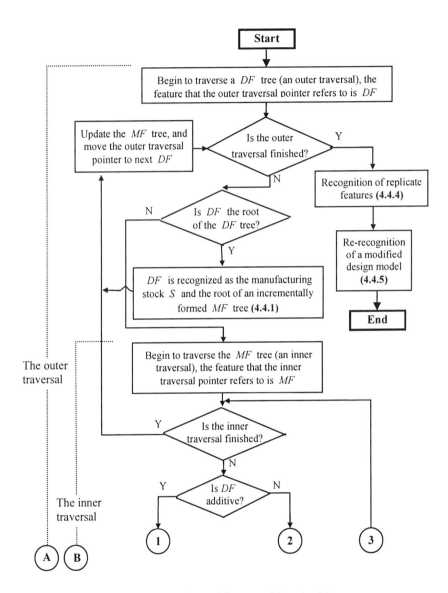

Fig. 4.6 Flowchart of the recognition algorithm.

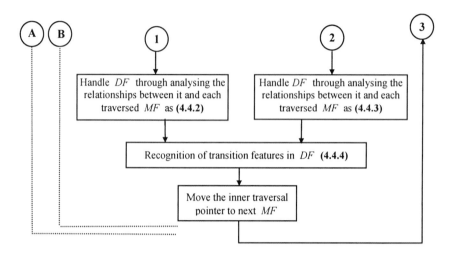

Fig. 4.6 Flowchart of the recognition algorithm (cont'd).

4.3.1 *Recognition of initial manufacturing stock*

The first traversed *DF* in the outer traversal operation is the root of the *DF* tree. It is directly recognised to form the root of an incrementally formed *MF* tree, i.e., manufacturing stock *S*. The volume of *S* can be increased when a new *DF* is recognised later. The outer traversal operation will continue to reach a new *DF* in the *DF* tree.

4.3.2 *Recognition of DF when it is additive*

For the currently traversed *DF*, the complete cycle of an inner traversal operation on the incrementally formed *MF* tree will be performed once, and the interacting relationships between *DF* and each *MF* traversed in the *MF* tree will be analysed and handled according to the following conditions (the detailed process is illustrated in Fig. 4.7, in which (1) or (2) means Steps (1) or (2) in the following process):

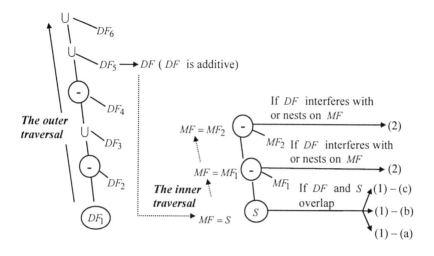

(a) *A design feature tree*   (b) *A MF tree after a DF is recognised*

Fig. 4.7 The traversal processes for recognising an additive design feature.

(1) If the traversed MF is S, and DF and S (i.e., MF) interact, there are three types of relationships between the extended directions of DF and S, which are denoted as $\overrightarrow{D_S}$ and $\overrightarrow{D_{DF}}$ respectively. These relationships include: (a) $\overrightarrow{D_S} \times \overrightarrow{D_{DF}} = 0$ (which means $\overrightarrow{D_S}$ and $\overrightarrow{D_{DF}}$ are parallel); (b) $\overrightarrow{D_S} \bullet \overrightarrow{D_{DF}} = 0$ (which means $\overrightarrow{D_S}$ and $\overrightarrow{D_{DF}}$ are perpendicular); and (c) $\overrightarrow{D_S} \times \overrightarrow{D_{DF}} \neq 0$ and $\overrightarrow{D_S} \bullet \overrightarrow{D_{DF}} \neq 0$ (which means there are not parallel or perpendicular relationships between $\overrightarrow{D_S}$ and $\overrightarrow{D_{DF}}$). Each relationship is processed according to one of the following steps respectively (during the process, a plane that is perpendicular to $\overrightarrow{D_S}$ is defined, and the boundaries of DF and S are projected onto it to form two areas, namely, F_{DF} and F_S):

(a) $\overrightarrow{D_S} \times \overrightarrow{D_{DF}} = 0$. There are two possible branches for this relationship:

[Case 1: $F_{DF} \subset F_S$] A base face of S towards DF is extended along $\overrightarrow{D_S}$ or its reverse direction until DF is covered completely. The extended volume is V_1. $V_1 - DF$ joins the MF tree as a new feature(s)[†] with an extended direction $\overrightarrow{D_S}$. S is updated as $S = S \cup V_1$.

[Case 2: $F_{DF} \not\subset F_S$] The maximum and minimum projected points of the boundary faces of DF and S along $\overrightarrow{D_S}$ are calculated. A base face of S is enlarged as $F_{DF} \cup F_S$, and it is extended along $\overrightarrow{D_S}$ and its reverse direction respectively, until it reaches the maximum and minimum projected points. The extended volume is V_1. $V_1 - DF$ joins the MF tree as a new feature(s) with an extended direction $\overrightarrow{D_S}$. S is updated as $S = S \cup V_1$.

(b) $\overrightarrow{D_S} \bullet \overrightarrow{D_{DF}} = 0$. There are two possible branches for this relationship:

[Case 1: $F_{DF} \subset F_S$] First, a base face of S towards DF is extended along $\overrightarrow{D_S}$ or its reverse direction until DF is covered completely. The extended volume is V_1. Next, a base face of DF is extended along $\overrightarrow{D_{DF}}$ and its reverse direction until they meet the boundaries of V_1. The extended volume is V_2. $V_1 - DF - V_2$ and $V_2 - DF$ join the MF tree as new features. The extended directions of $V_1 - DF - V_2$ and

---

[†] $V_1 - DF$ might consist of one or several separated partitions. Each partition will be recognised as a manufacturing feature. The following $V_1 - DF - V_2$, $V_2 - DF$, etc., are handled similarly.

$V_2 - DF$ are $\overrightarrow{D_S}$ and $\overrightarrow{D_{DF}}$ respectively. S is updated as $S = S \cup V_1$.

[Case 2: $F_{DF} \not\subset F_S$] First, the maximum and minimum projected points of DF and S along $\overrightarrow{D_S}$ are calculated. A base face of S is enlarged as $F_{DF} \cup F_S$, and it is extended along $\overrightarrow{D_S}$ or its reverse direction until it reaches the maximum and minimum projected points. The extended volume is V_1 . Next, the base faces of DF are extended along $\overrightarrow{D_{DF}}$ until they meet the boundary of V_1 . The extended volume is V_2 . $V_1 - DF - V_2$ and $V_2 - DF$ join the MF tree as new features. The extended directions of $V_1 - DF - V_2$ and $V_2 - DF$ are $\overrightarrow{D_S}$ and $\overrightarrow{D_{DF}}$ respectively. S is updated as $S = S \cup V_1$.

(c) $\overrightarrow{D_S} \times \overrightarrow{D_{DF}} \neq 0$ and $\overrightarrow{D_S} \bullet \overrightarrow{D_{DF}} \neq 0$. There are two possible branches for this relationship:

[Case 1: $F_{DF} \subset F_S$] First, a base face of S towards DF is extended along $\overrightarrow{D_S}$ or its reverse direction until DF is covered completely. The extended volume is V_1 . Next, a base face of DF out of S is enlarged to divide V_1 into two partitions, V_{11} and V_{12} . $V_{11} - DF$ and $V_{12} - DF$ join the MF tree as new features with extended directions $\overrightarrow{D_S}$. S is updated as $S = S \cup V_1$.

[Case 2: $F_{DF} \not\subset F_S$] First, the maximum and minimum projected points of S and DF along $\overrightarrow{D_S}$ are calculated. A base face of S is enlarged to cover the boundary of $F_{DF} - F_S$, and it is extended along $\overrightarrow{D_S}$ and its reverse direction respectively, until it reaches the maximum and minimum projected points.

The extended volume is V_1. Next, a base face of DF out of S is enlarged to divide V_1 into two partitions, V_{11} and V_{12}. $V_{11} - DF$ and $V_{12} - DF$ join the MF tree as new features with extended directions $\overrightarrow{D_S}$. S is updated as $S = S \cup V_1$.

(2) If DF interferes with or nests on MF, MF is updated as $MF - DF$.

For the part shown in Fig. 4.8, DF_1, which is the first traversed design feature, is recognised as the initial S of an incrementally updated MF tree. For DF_2, the current MF tree has only one feature S. The relationship between DF_2 and S falls in the above Step $(1) - (a) - (i)$. S is lifted along $\overrightarrow{D_S}$ to generate V_1. $MF_1 = V_1 - DF_2$ is generated. S is updated as $S = S \cup V_1$. The MF tree is updated to consist of S and MF_1.

For the part shown in Fig. 4.9, the relationship between DF_2 and S, which has been recognised from DF_1, falls in the above Step $(1) - (a) -$ (ii). A base face of S is enlarged as $F_{DF_2} \cup F_S$ and extended along $\overrightarrow{D_S}$ and its reverse direction respectively, until it reaches the maximum and minimum projected points. $V_1 - DF$ has two partitions and these partitions join the MF tree as new features. S is updated as $S = S \cup V_1$.

The recognition processes of DF_2 in the parts shown in Fig. 4.10 to Fig. 4.13 fall in the above Steps $(1) - (b) - (i)$, $(1) - (b) - (ii)$, $(1) - (c) -$ (i), $(1) - (c) - (ii)$, respectively. Their processes are illustrated in these figures.

In Fig. 4.14, the additives DF_3 and DF_4 nest on MF_3 and MF_4, which have been recognised from DF_2. This case can be handled as the above Step (2). MF_3 and MF_4 are updated as $MF_3 = MF_3 - DF_3$ and $MF_4 = MF_4 - DF_4$.

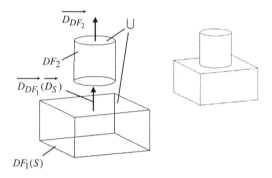

(a) *An additive* DF_2 *will be recognised*

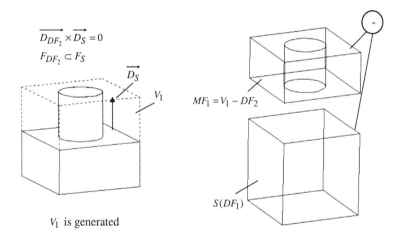

(b) *The recognition process of* DF_2 (c) *The MF tree after* DF_2 *is recognised*

Fig. 4.8 The recognition process of an additive feature that satisfies (1) – (a) – (i).

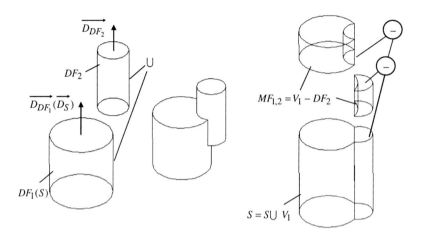

(a) *An additive DF_2 will be recognised*   (b) *The MF tree after DF_2 is recognised*

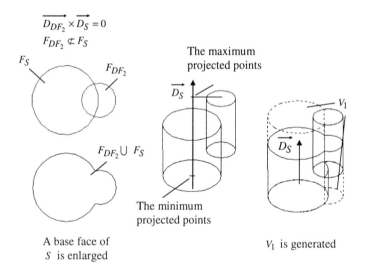

(c) *The recognition process of DF_2*

Fig. 4.9 The recognition process of an additive feature that satisfies (1) – (a) – (ii).

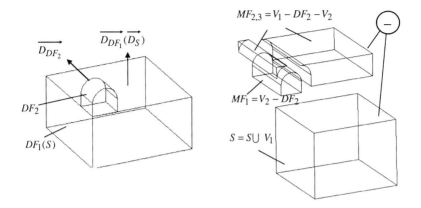

(a) *An additive DF_2 will be recognised*   (b) *The MF tree after DF_2 is recognised*

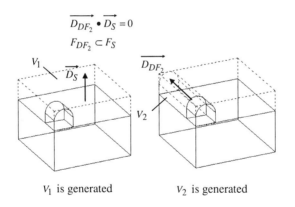

$$\overrightarrow{D_{DF_2}} \bullet \overrightarrow{D_S} = 0$$
$$F_{DF_2} \subset F_S$$

V_1 is generated V_2 is generated

(c) *The recognition process of DF_2*

Fig. 4.10 The recognition process of an additive feature that satisfies (1) – (b) – (i).

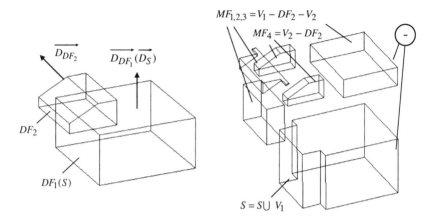

(a) *An additive DF_2 will be recognised* (b) *The MF tree after DF_2 is recognised*

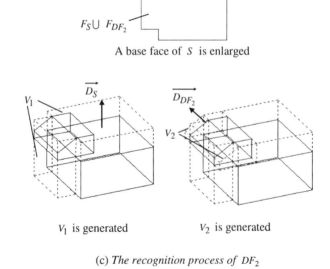

(c) *The recognition process of DF_2*

Fig. 4.11 The recognition process of an additive feature that satisfies (1) – (b) – (ii).

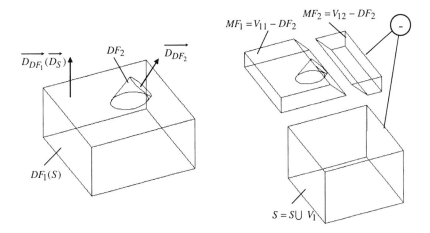

(a) *An additive DF_2 will be recognised*   (b) *The MF tree after DF_2 is recognised*

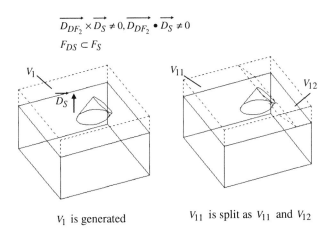

V_1 is generated V_{11} is split as V_{11} and V_{12}

(c) *The recognition process of DF_2*

Fig. 4.12 The recognition process of an additive feature that satisfies (1) – (c) – (i).

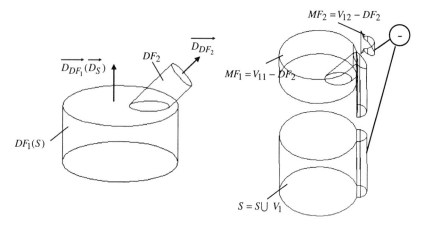

(a) *An additive* DF_2 *will be recognised*   (b) *The MF tree after* DF_2 *is recognised*

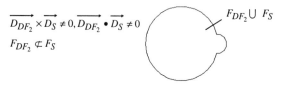

A base face of S is enlarged

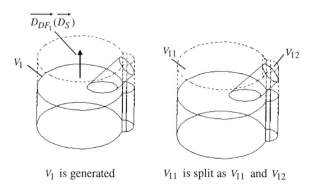

V_1 is generated V_{11} is split as V_{11} and V_{12}

(c) *The recognition process of* DF_2

Fig. 4.13 The recognition process of an additive feature that satisfies (1) – (c) – (ii).

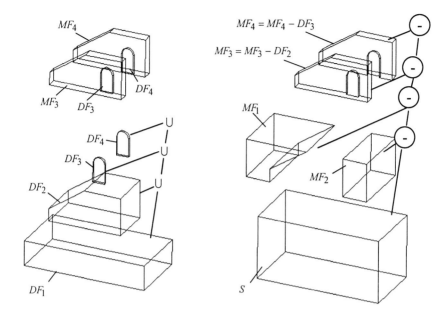

(a) *A design feature tree*   (b) *The MF tree after DF_3 and DF_4 are recognised*

Fig. 4.14 The recognition process of additive features that satisfies (2).

4.3.3 *Recognition of DF when it is subtractive*

DF is handled through analysing its relationships with each MF traversed in the MF tree starting from the root to the highest leaf according to the following conditions:

(1) If MF is S, and DF penetrates S, DF is updated as $DF \cap S$.

(2) If DF and MF satisfy any case listed in Table 4.3, they are handled as below:

- In cases (a), (c), (g), and (i), DF and MF are kept the same.
- In cases (b), (e), and (h), DF is updated as $DF - MF$, and MF is kept the same.
- In cases (d) and (f), MF is updated as $MF - DF$, and DF is kept the same.

(3) *DF* and *MF* are kept the same if the other interactions occur between them.

DF might satisfy several of the above conditions with different *MF* being traversed. The finally updated *DF* joins the *MF* tree as a new feature. A sketch of the above process is shown in Fig. 4.15.

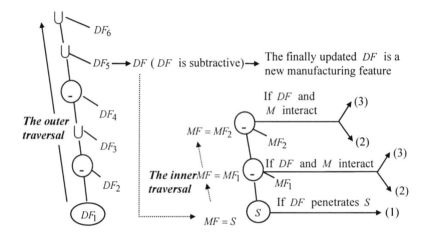

(a) *The design feature tree*   (b) *The MF tree before DF is recognised*

Fig. 4.15 The recognition process of a subtractive design feature.

Table 4.3 The mutual interacting relationships between *MF* and *DF* in the adjacency, splitting and disconnection categories.

Relationships	*MF* is adjacent	*MF* is split	*MF* is disconnected
DF is adjacent	(a)	(d)	(g)
DF is split	(b)	(e)	(h)
DF is disconnected	(c)	(f)	(i)

A part in Fig. 4.16 is used to illustrate several of these cases. DF_1 is directly converted to S. Since DF_2 penetrates S, $MF_1 = DF_2 \cap S$ is generated and join the current MF tree. Similarly, DF_3 penetrates S, and it is updated as $DF_3 \cap S$. Since the relationship between the updated DF_3 and MF_1 satisfies (2) – (b), $MF_2 = (DF_3 \cap S) - MF_1$ is generated and joins the MF tree, and MF_1 is kept the same. The final MF tree consists of S, MF_1 and MF_2.

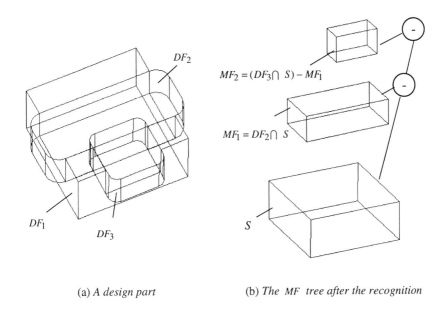

(a) *A design part*   (b) *The MF tree after the recognition*

Fig. 4.16 The recognition process of subtractive features.

4.3.4 *Recognition of auxiliary and replicate features*

DF might have an auxiliary design feature, which is usually a local operation on an edge or vertex, and uses a single face to connect the adjacent faces of the edge or vertex. The face of an auxiliary feature can be a planar face or a surface, and the edge or vertex operated on can be convex or concave in the part. The convexities of an edge or vertex are given below:

- An edge in a part is convex if the angle between its two adjacent faces in the part is between 180° and 360°; otherwise, the edge is concave.
- A vertex in a part is convex if the edges from it in the part are convex; otherwise, the vertex is concave.

An auxiliary *DF* is handled in one of the following four ways:

(1) If the face of the auxiliary feature is planar, and the edge or vertex of the feature is convex, the auxiliary feature is handled as a subtractive feature as in Section (4.4.3). The feature that is recognised from it is a single feature, namely, a chamfer.

(2) If the face is planar, and the edge or vertex is concave, the auxiliary feature is handled as an additive feature as in Section (4.4.2), and the feature recognised from it is a single feature, namely, a fillet.

(3) If the face is a surface, and the edge or vertex is convex, the auxiliary feature is handled as a subtractive feature as in Section (4.4.3), and the feature recognised from it is a single feature, namely, a round edge feature.

(4) If the face is a surface, and the edge or vertex is concave, the auxiliary feature is handled as an additive feature as in Section (4.4.2), and the feature recognised from it is a single feature, namely, a fillet.

Fig. 4.17 shows the recognition of a part with two auxiliary design features. DF' joins MF_1 as a fillet. DF'' is recognised as MF_2 - a round edge feature.

In a *MF* tree, several features can be grouped together into a new replicate feature if they form certain pattern relationships and have the same (TAD, the tool position face, the cutting depth).

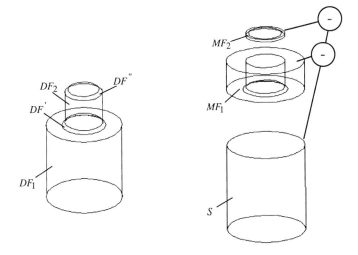

(a) *A part with two auxiliary design features*   (b) *The MF tree for the part*

Fig. 4.17 The recognition process of auxiliary design features.

4.3.5 Re-recognition of modified design model

A designer can interactively modify the part to remove manufacturing interactions caused by a poor design or modify some unsatisfied details. The modification operations include: addition of a new design feature, deletion of a design feature, and edition of a design feature. The recognition processor will handle the modified part according to the following steps:

(1) Addition of a new design feature. The new feature is handled as a new feature added to the *MF* tree. The Steps from Section (4.4.2) – Section (4.4.4) are triggered to update the *MF* tree.

(2) Deletion of a design feature. There are three conditions: (a) the feature is subtractive; (b) the feature is additive and there is no corresponding manufacturing feature generated from it; and (c) the feature is additive and it has a corresponding manufacturing feature

in the intermediate tree. The recognition process will be carried out according to one of the following processes:

(a) The corresponding manufacturing feature of the design feature in the *MF* tree is deleted. The shrunk volumes of the manufacturing features in the *MF* tree, which are caused by the interacting relationships with this design feature, should be restored. The other features in the *MF* tree are kept the same.

(b) The shrunk volumes of the manufacturing features in the *MF* tree, which are caused by the interference or nesting relationships with this design feature, should be restored. The other features in the *MF* tree are kept the same.

(c) The *MF* tree before this feature is recognised and restored. The recognition process restarts from this point for the updated design feature tree.

(3) Edition of a design feature. An edition operation can be handled as two consecutive steps: (i) delete the old feature, and (ii) add the new feature. Hence, this operation can be combined by the above steps (1) and (2) consecutively.

Fig 4.18 shows the recognition of a part [Wang and Kim, 1998] according to the processes presented in Sections (4.4.1)-(4.4.4). The part and the recognised *MF* tree are shown in Fig. 4.18, in which MF_6, MF_7 and MF_8 are replicated features. The modifications of a part are shown in Fig. 4.19. In Fig. 4.19, the subtractive DF_9 is deleted according to the above Condition (2) – (a). MF_6 and MF_7 interact with DF_9 during the recognition process. Hence, the corresponding shrunk volumes of MF_8 and MF_9 caused by DF_9 are resumed. The other features in the *MF* tree are kept the same. In Fig. 4.19, the additive DF_7 is deleted according to the above condition (2) – (b) since there is no corresponding recognisable manufacturing feature. The shrunk volume of MF_5 caused by the nesting of DF_7 should be restored. The other features in the *MF* tree are kept the same. In Fig. 4.19, the additive DF_2 is deleted. Since MF_1 is recognised from DF_1, according to the above condition (2) – (c), the recognition process restarts from recognising DF_3.

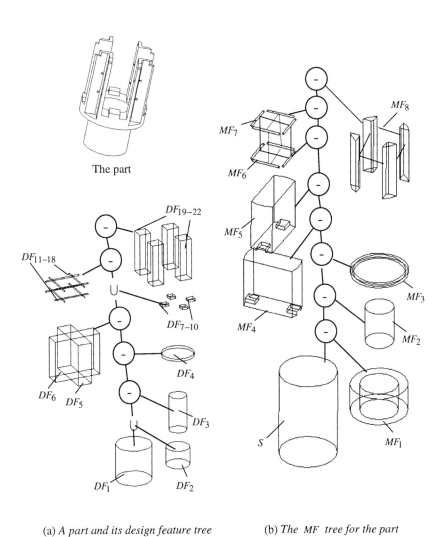

The part

DF_{19-22}

DF_{11-18}

DF_{7-10}

DF_4

DF_6 DF_5

DF_3

DF_1 DF_2

MF_7

MF_8

MF_6

MF_5

MF_4

MF_3

MF_2

S

MF_1

(a) *A part and its design feature tree*   (b) *The MF tree for the part*

Fig. 4.18 Kim 2 part and its recognised *MF* tree.

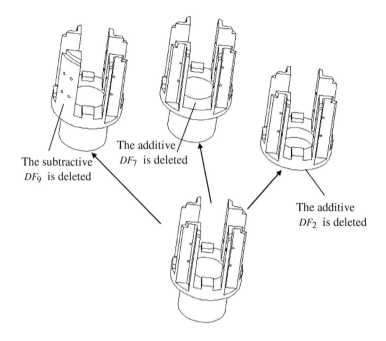

Fig. 4.19 The part is modified.

4.4 Multiple Manufacturing Feature Interpretations

4.4.1 *Properties of manufacturing feature tree*

A *MF* tree stores a single interpretation of the manufacturing features, and it can be expressed as:

$$P = S - MF_1 - ... - MF_n \qquad (4.1)$$

where *P* is the design part, *S* and MF_i are the recognised manufacturing stock and feature respectively.

The Equation (4.1) and the features in Equation (4.1) have the following properties:

(1) Associativity. In the Equation (4.1), if several features are combined as a single feature, the geometry of the Equation (4.1) is kept the same. For instance, if

$$MF' = MF_i \cup MF_j \tag{4.2}$$

and assume

$$R = S - MF_1 - \ldots \; MF_{i-1} - MF_{i+1} - \ldots - MF_{j-1} - MF_{j+1} - \ldots - MF_n \tag{4.3}$$

then

$$P = S - MF_1 - \ldots - MF_n = R - MF_i - MF_j = R - MF' \tag{4.4}$$

(2) Interchangeability. In the Equation (4.1), if the sequences of any two manufacturing features are interchanged, the geometry of Equation (4.1) is kept the same. For instance, if MF_i and MF_j in Equation (4.1) are interchanged, then

$$P = S - MF_1 .. - MF_i - \ldots - MF_j - \ldots - MF_n = S - MF_1 \ldots - MF_j - \ldots - MF_i - \ldots - MF_n \tag{4.5}$$

(3) Replacement. In Equation (4.1), if a feature is equal to several combined features, when the feature is replaced with these features, the geometry of Equation (4.1) is kept the same. For instance, if

$$MF_i = \cup_{k=1\ldots l} MF'_k \tag{4.6}$$

then,

$$P = S - MF_1 - \ldots - MF_i - \ldots - MF_n = S - MF_1 - \ldots - \cup_{k=1\ldots l} MF'_k - \ldots - MF_n \tag{4.7}$$

(4) Enclosure. For each feature in Equation (4.1), its volume is enclosed in the manufacturing stock. For instance, any feature MF_i is

$$MF_i \subset S \tag{4.8}$$

The faces of a manufacturing feature in a MF tree can be classified into three types, namely, faces that coincide with the faces of the manufacturing stock S (type I), faces that coincide with the faces of the part (type II), and the remaining faces (type III). A type I or III face in a feature can be used to position a cutting tool, and the normal vector of the face towards the feature is a *possible* TAD of the feature. A type II face cannot be intruded and sets the boundary constraint for the movement of a cutting tool. A feature can extend its volume outwards along the normal of a type III face until the extended part meets a type I or II face of another feature. The extended volume is the optional volume of the feature. Fig. 4.20 shows these concepts.

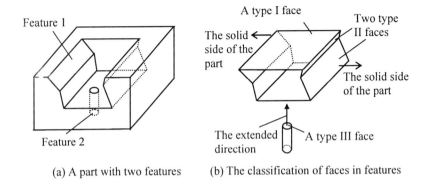

(a) A part with two features   (b) The classification of faces in features

Fig. 4.20 The classification of faces in manufacturing features.

4.4.2 *Generation of multiple interpretations*

Based on these properties and concepts, the recognised *MF* tree can be extended with AND/OR operators to store the multiple interpretations. Three geometric operations, i.e., combination, decomposition, and TAD-led operation, are used to generate the multiple interpretations.

4.4.2.1 Combination

The following four conditions can generate an alternative interpretation through various combinations:

(1) If two features in an interpretation have a virtually linked interaction, a new interpretation is generated through replacing them with a united feature.

(2) In an interpretation, if several holes abut and are co-axial, these holes are united together as a new combined feature. A new interpretation is generated through replacing the former features with this combined feature.

(3) In an interpretation, if a slot abuts other slots, and these slots are co-symmetrical and have the same TADs, these slots can be united together as a new compound feature. A new interpretation is

generated through replacing the former features with this compound feature.

(4) If two features are abutted/adjacent/disconnected in an interpretation, and they can be machined with the same set of TAD, tool position face and the cutting depth, a new interpretation is generated through replacing them with a united feature.

The features that satisfy the above conditions in the *MF* tree are split from the *MF* tree, and an AND operator is used to link these features as children of this *MF* tree. An OR operator, which is used to link this AND operator and the single combined feature from these features, is re-connected to the top of the tree via a Boolean difference operator. The AND operator means if one of its children is chosen in an interpretation, the others must also be included. The OR operator implies that its children are mutually exclusive and each interpretation can only include one of them.

In Fig. 4.21, MF_1 and MF_2 are two abutted and co-symmetrical slots, and they have the same TADs. Hence, they can be combined as a compound slot.

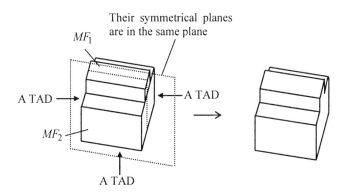

Fig. 4.21 MF_1 and MF_2 can be combined as a compound slot.

4.4.2.2 Decomposition

The following conditions can generate alternative interpretations through decomposition:

(1) From the extended direction of a feature MF_i, if there are type II faces inside MF_i perpendicular to this direction, these faces are extended until they meet the boundary of MF_i. This boundary is divided into several new features $MF'_{1...k}$ by these extended faces;

(2) If a pair of opposite parallel faces with a common adjacent face in a feature MF_i can be extended until they meet the boundary of the manufacturing stock S without intruding the other faces in the part, a slot MF'_i can be obtained. A new interpretation can be generated by replacing MF_i with MF'_i and $MF_i - MF'_i$.

Several decomposed features from MF_i are linked with an AND operator. An OR operator, which is used to link this AND operator and MF_i, is re-connected to the previous parent node of MF_i. In Fig. 4.22, from the extended direction of a feature MF, there are two type II faces inside MF perpendicular to the TAD. These faces are extended to divide MF into three features MF'_1, MF'_2 and MF'_3.

4.4.2.3 TAD-led operations

If there is a type II face or an edge between two type II faces in a feature, and along one of the possible TADs in the feature, a straight line passes through the face twice or both of the adjacent faces of the edge, or the face cannot be approached, this TAD is inaccessible and deleted. Otherwise, this possible TAD is considered feasible. A type I or III face for the TAD in the feature is the tool position face. The cutting depth is the maximum distance for a tool to move along a TAD from its tool position face. In Fig. 4.23, an inaccessible TAD, and a feasible TAD, as well as its tool position face and cutting depth are illustrated. If a feature has several TADs, the feature can be machined from several directions, and therefore, alternative machining plans can be generated. In the MF tree, TADs for each feature are labelled in the Cartesian coordinates.

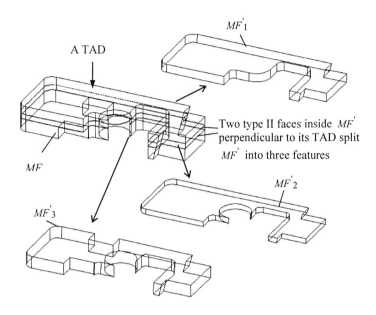

Fig. 4.22 The decomposition operations on a feature.

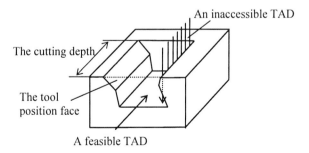

Fig. 4.23 An inaccessible TAD, a feasible TAD and its machining elements.

The recognition of the ANC-101 part from [Shah, *et al.*, 1994], with its *MF* tree extended with AND/OR operators to store the multiple interpretations, is shown in Fig. 4.24.

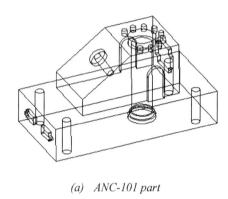

(a)   ANC-101 part

Fig. 4.24 The ANC-101 part and its *MF* tree extended with AND/OR operators.

4.4.3 *Optimal single interpretation of manufacturing features*

In a specific workshop environment, different interpretations of features in a manufacturing feature tree are associated with different costs, and the lowest machining cost will be selected for process planning. A processor will choose the interpretation depending on the machining cost of features in the manufacturing feature tree. However, the selection of the TADs for the features can be quite complicated and is dependent on the optimal set-up and operations sequence. Hence, the features in a selected interpretation are kept in relation to all its TADs, and a chosen interpretation could consist of several machining plans.

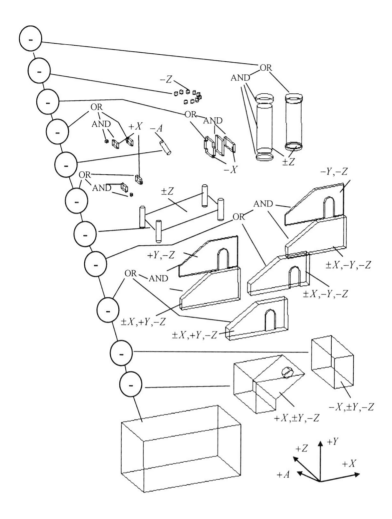

(b) *The updated manufacturing feature tree with AND/OR operators*

Fig. 4.24 The ANC-101 part and its *MF* tree extended with AND/OR operators (cont'd).

The cost of a feature or an AND/OR operator in a MF tree is expressed as $Cost(MF)$ or $Cost(AND/OR)$. The total cost is $Total\_Cost$, and for the selected interpretation is initially set to 0. The selection and cost calculation process is described in the following steps:

(1) Set $Total\_Cost = Cost(S)$, where S is the manufacturing stock.

(2) Traverse the Boolean operator in the MF tree from the operator nearest to the root to the highest operator. The process will follow one of the following two branches:

    (a) If the traversed Boolean operator is connected with a feature MF_i , $Total\_Cost = Total\_Cost + Cost(MF_i)$, and the feature together with its TAD is chosen as an element in the interpretation.

    (b) If the traversed Boolean operator is connected with an OR operator (denoted as OR_TOP in the following content), the cost of its sub-tree is computed from the lowest level AND/OR operator backwards to the OR_TOP . Each AND/OR operator is assumed to have L sub-AND/OR operators and M single features, i.e., $AND/OR_{i=1...L}$ and $MF_{j=1...M}$, their costs are computed as follows:

        (i)   An AND operator:

$$Cost(AND) = \sum_{i=1}^{L} Cost(AND/OR_i) + \sum_{j=1}^{M} Cost(MF_j).$$

        (ii)   An OR operator:

$$Cost(OR) = Min \begin{cases} \sum\limits_{i=1}^{L} Min\,Cost(AND/OR_i) \\ \sum\limits_{j=1}^{M} Cost(MF_j) \end{cases}.$$

(c) $Total\_Cost = Total\_Cost + Cost(OR\_TOP)$. The features in the sub-tree contributing to the $Total\_Cost$ are chosen as the elements of the interpretation.

Fig. 4.25 shows an illustrative case to compute a sub-tree of a manufacturing feature tree.

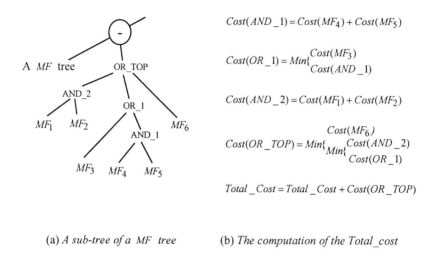

$$Cost(AND\_1) = Cost(MF_4) + Cost(MF_5)$$

$$Cost(OR\_1) = Min\{ \begin{matrix} Cost(MF_3) \\ Cost(AND\_1) \end{matrix}$$

$$Cost(AND\_2) = Cost(MF_1) + Cost(MF_2)$$

$$Cost(OR\_TOP) = Min\{ \begin{matrix} Cost(MF_6) \\ Min\{ \begin{matrix} Cost(AND\_2) \\ Cost(OR\_1) \end{matrix} \end{matrix}$$

$$Total\_Cost = Total\_Cost + Cost(OR\_TOP)$$

(a) *A sub-tree of a MF tree*   (b) *The computation of the Total_cost*

Fig. 4.25 The cost computation for a sub-tree of a *MF* tree.

4.5 Summary

Manufacturing applications typically require that a part created in a design-by-feature system be described through manufacturing features defined in STEP AP 224. This chapter presents a manufacturing feature recognition processor that utilises the high-level information retrieved from a design-by-feature system to generate manufacturing feature model efficiently. Through systematic definitions and processes, the interacting relationships between features can be represented and handled. By utilising the previous intermediate recognition information, a re-recognition processor is developed to dynamically recognise features from an updated design part after modification. From the manufacturing feature tree generated by the feature recognition processor described, through combination, decomposition, and TAD-led operations, alternative interpretations for a design part can be generated and stored in a manufacturing feature tree extended with AND/OR operators, which are used to support a generic and multiple-view manufacturing application environment. A simple comparison for the approach and other related research is shown in Table 4.4.

In summary, some characteristics of this approach are as follows:

The proposed processor recognises the manufacturing features from the design feature tree in a step-by-step manner according to the evolving process of a design plan. The processor and a design-by-feature system can be further integrated in a concurrent engineering environment, so that the processor and the manufacturing evaluation can be triggered to evaluate the design plan dynamically and efficiently when the design plan evolves.

Using a systematic analysis, the interacting relationships between features can be represented and handled elaborately. With the system integration based on STEP, the manufacturing feature models can be shared by other manufacturing applications in an internationally accepted standard. The re-recognition process can facilitate the recognition process to update a modified part so that a designer can modify some manufacturing defects and attempt alternative design plans.

Based on the relationships of feature geometric relationships, the efficiency of the processor for generating alternative interpretations from a single interpretation is high compared with the existing approaches from geometric modelling. The generated alternative interpretations contain the same or similar sets of manufacturing features from different constructive models of the same part. The AND/OR tree provides a concise data structure to represent the alternative interpretations. Through assigning costs for the different features in the AND/OR tree in a specific environment, a reasonable interpretation with the lowest cost can be retrieved easily.

Table 4.4 A comparison of several feature recognition approaches.

Research	Capability for Recognising Interacting Features	Multiple Interpretations
Laakko and Mantyla, 1993	Weak	No
Han and Requicha, 1997	Able to recognise features from some hints of interacting relationships, such as parallel and pocket wall faces.	Yes
Suh and Ahluwalia, 1995	Facilitate the re-recognition process of certain types of interacting features, such as adjacency, splitting and disconnection [*].	No
Perng and Chang, 1997	Facilitate the re-recognition process for the modification of some interacting features. The interacting relationships include nesting, interference, penetration, and overlapping [*].	No
Lee and Kim, 1998, 1999	Able to recognise certain types of interacting features, such as nesting, interference, and overlapping [*].	Yes
This approach	Able to recognise most types of interacting features.	Yes

[*] The terminology of the interacting relationships is different from that of the research.

Chapter 5

Intelligent Optimisation of Process Planning

CAPP is an essential component for linking design and downstream manufacturing processes. It has been motivated by the desire to free humans from planning manufacturing processes for design parts and to optimise decisions required during these processes, so as to achieve a greater uniformity in the product development life-cycle. A CAPP system should ideally generate optimised process plans to ensure the application of good manufacturing practices and maintain the consistency of the desired functional specifications of a part during its production processes. Crucial processes in a CAPP system, such as selecting machining resources, determining set-up plans and sequencing operations of a prismatic part, should be considered simultaneously to achieve global optimal solutions. In this chapter, these processes are integrated and modelled as a constraint-based optimisation problem, and intelligent heuristic search-based approaches are presented to solve it effectively.

5.1 Intelligent Optimisation Strategies for CAPP Systems

CAPP systems that have been reported can be classified into two categories, namely, the variant systems and generative systems.

The initial variant CAPP systems are based on the Group Technology (GT) to select a baseline process plan for a part family. The principle justifying such systems is that design parts with similar geometrical shapes and technical requirements have similar process plans. The essential component in a variant system is a repository of process plans edited by experienced planners through retrieving the relevant data for similar products and organising them using GT coding and classification approaches. For a new part, a variant system can assist to identify similar plans and edit them according to the geometrical and technological differences. Approximately 90% of the part attributes and parameters can be yielded automatically while the remaining 10% is achieved through modifying or fine-turning the process plans manually.

Current systems are moving towards generative systems and AI techniques are used to facilitate the generation of a complete process plan for a new part from scratch. In a generative CAPP system, knowledge and expertise about manufacturing processes are encoded to support automatic decision logic instead of retrieving and editing an existing process plan through intervention of a planner as in a variant system. An important trend in generative CAPP systems is their adaptability towards dynamically varying manufacturing conditions, such as the availability of machines and tools. Conventionally, a generative CAPP system consists of three main consecutively activities, namely, (1) recognising manufacturing features from a design part, (2) determining machining operation types and enumerating alternative set-up plans as well as applicable machining resources in a dynamic workshop environment, and (3) selecting suitable set-up plans and machining resources, and sequencing machining operations to seek the lowest machining cost of the part. The workflow of a generative CAPP system is shown in Fig. 5.1.

A global optimum process plan can be achieved through an optimisation of each individual activity. The third activity, which is the focus here, can be modelled as an optimisation problem and solved using AI techniques. However, such a problem is well known to be an intractable reasoning and decision-making process considering the inter-related geometric relationships between features, the complex technological requirements and the multiple evaluation criteria. To address this problem effectively, efforts should be made to design a more apt optimisation model and develop a more efficient method for handling precedence constraints of a part.

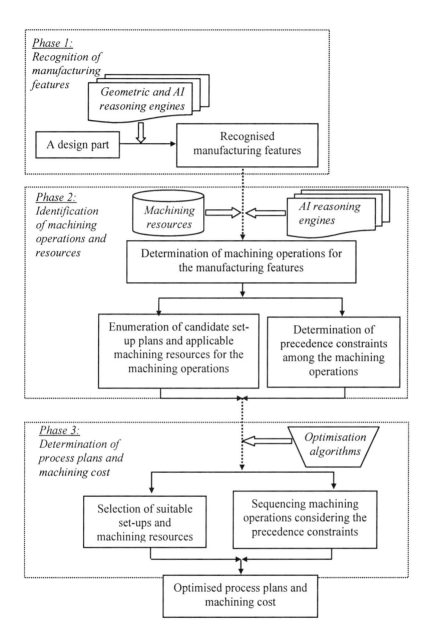

Fig. 5.1 The workflow of a generative CAPP system.

Recent related works can be categorised into the knowledge-based reasoning approach [Chang, 1990; Wong and Siu, 1995; Chu and Gadh, 1996; Wu and Chang, 1998; Tseng and Liu, 2001], graph manipulation approach [Chen and LeClair, 1994; Irani *et al.*, 1995; Lin *et al.*, 1998], Petri-nets-based approach [Kruth and Detand, 1992], fuzzy logic reasoning approach [Zhang and Huang, 1994; Ong and Nee, 1994; Gu *et al.*, 1997], evolutional algorithm and heuristic reasoning-based approach [Vancza and Markus, 1991; Chen and LeClair, 1994; Yip-Hoi and Dutta, 1996; Zhang, *et al.*, 1997; Chen, *et al.*, 1998; Reddy, *et al.*, 1999; Qiao, *et al.*, 2000; Ma, *et al.*, 2000; Lee, *et al.*, 2001], etc. In the reasoning processes, the methodology to manipulate the preliminary precedence constraints between operations effectively, which can be represented as "if-then" rules, graphs and matrices, is an important issue.

The QTC system developed by Chang [1990] is a knowledge-based reasoning approach that aggregates machining operations with the same TAD to form a set-up. The sequence of the machining operations and set-ups is reasoned according to the precedence constraints, and an optimal sequence is selected from several feasible sequences based on the minimum number of set-ups. In this system, several geometric and technological constraints of a part, stemming from geometric interactions between operations, location tolerance requirements, reference or datum requirements, and good manufacturing practices, are considered in the reasoning process. The aggregation concept is enhanced by Chua and Gadh [1996] to cluster operations that are machined with the same cutting tool into a set-up so as to reduce the number of tool changes. In the APSS system reported by Wong and Siu [1995], the operations sequencing algorithm consists of three consecutive algorithms, viz., the transformation algorithm, the refinement algorithm, and the linearisation algorithm. After the preliminary precedence constraints have been generated in the transformation algorithm, the refinement algorithm creates the details of the operations using a "refinement" knowledge base considering good manufacturing practices, and represent them in a tree structure. In the linearisation algorithm, the tree structure will be linearised into the final required operation sequence.

Irani, *et al.* [1995] developed a graph manipulation approach for operations sequencing, in which the Hamiltonian Path (HP) analogy is utilised to reason process plans and the Latin Multiplication Method (LMM) is implemented to enumerate all the feasible HPs under constraints. The optimal process plan is an HP that corresponds to the least number of set-up disruptions required from start to finish to process each feature once and only once. In the work reported by Lin *et al.* [1998], a graph-search strategy was designed for operations sequence planning for prismatic parts with interacting features. The machining of a feature might affect the surface quality of other interacting features, and several heuristic rules concerning machining practices were developed to help the graph-search process generate high quality process plans with the lowest machining cost. A hybrid method combining an unsupervised learning algorithm and a graph manipulation algorithm was proposed by Chen and LeClair [1994]. The sequence of the features in each set-up is constrained and determined by the interacting feature relationships, which are stored in an episodal associative memory. After the process of feature sequencing, several graph manipulation algorithms are proposed to obtain the optimal tools sequence for creating the features in a set-up by minimising the tool changes.

Kruth and Detand [1992] proposed generic Petri-nets to represent parametric features and their related machining operations. After the features have been evaluated using the manufacturing knowledge bases, such as general machine data and manufacturing capability data, the separated Petri-nets for compound features or features with identical TADs are first joined together. The same procedure is then applied to the features with different TADs, and a large Petri-net is finally formed, in which all the valid alternatives to machine the part are described.

The common characteristics of the above approaches include: (1) although these reasoning approaches can generate feasible solutions, it is very difficult to find the global and optimal plans using these approaches, and (2) the reasoning efficiency is low in a complex machining environment.

Process planning objectives are often imprecise and can even be conflicting due to inherent differences in the feature geometry and technological requirements. Fuzzy logic is suitable for presenting such imprecise knowledge and several approaches using this technique have been developed to address the process planning problem. Based on fuzzy memberships, the objective [Zhang and Huang 1994; Tiwari and Vidyarthi 1998] is to minimise the dissimilarity among the process plans selected for a family of parts, and optimal process plans can be generated for each part family. A fuzzy logic-based approach has been reported by Ong and Nee [1994] to identify and prioritise important features based on the geometric and technological information of a part. Important features and their operations correlate well with the manufacturing cost than the less important features and operations. Hence, operations sequencing of important features is first carried out within a much smaller search space. The operations of the less important features can then be arranged easily due to reduced constraints.

Recently, evolutional and heuristic algorithms based on the TS, SA, and GA techniques have been applied to the process planning research, and multiple objectives, such as the minimum use of expensive machines and tools, minimum number of set-ups, machine changes and tool changes, and achieving good manufacturing practice, have been incorporated and considered as a unified model to achieve a global optimal target. The related works are summarised in Table 5.1. The popular strategies for generating neighbourhood or next generation process plans in TS, SA, and GA are illustrated in Tables 5.2 and 5.3. However, there are two issues that are still outstanding and require careful considerations.

The first issue is that the multiple objectives in the unified optimisation models are often conflicting, and a prismatic part with alternative TADs and cutting machines/tools creates a large search space. Much effort is needed to design a suitable optimisation model to determine the optimal results with short reasoning iterations to meet the practical workshop situations with dynamically varying resources and workloads. It is also imperative to conduct comprehensive studies on the performance of the various optimisation algorithms to highlight their characteristics.

The second issue is the processing of precedence constraints in the various optimisation approaches. The efficiencies of the graph-based heuristic algorithm [Vancza and Markus, 1991] and the tree traversal algorithm [Yip-Hoi and Dutta, 1996] are not high and the search is not global, such that optimal plans might be lost during the inference and

Table 5.1 Summary of the related work for evolutional and heuristic algorithm-based process plan optimisation.

Related Works	Optimisation Strategies	Optimisation Criteria	Adjustment Strategies*	Constraint Types	Constraint Handling
Vancza and Markus, 1991	Genetic Algorithm	(1) Number of set-ups (2) Number of tool changes (3) Total cost of individual operations	(1) Crossover (2) Mutation	(1) Fixed order of operations (2) Reference precedence (3) Feature interactions	Graph-based heuristic algorithm
Yip-Hoi and Dutta, 1996	Genetic Algorithm	Part machining time	(1) Crossover (2) Mutation	(1) Fixed order of operations (2) Resource constraints	Tree traversal algorithm
Zhang, et al., 1997	Genetic Algorithm	(1) Machine costs (2) Cutting tool costs (3) Number of machine changes (4) Number of tool changes (5) Number of set-ups	(1) Crossover (2) Mutation	(1) Fixture constraints (2) Datum dependency (3) Fixed order of operations	(1) Test and generation (2) Two-step manipulation for crossover
Chen, et al., 1998	Hybrid Hopfield Neural Network and Simulated Annealing	(1) Number of set-ups (2) Number of tool changes (3) Number of constraint violation	Exchange of the rows in Neural Networks	(1) Fixture constraints (2) Feature interactions (3) Tolerance (4) Tool approach directions	Penalty algorithm

Table 5.1 Summary of the related work for evolutional and heuristic algorithm-based process plan optimisation (cont'd).

Related Works	Optimisation Strategies	Optimisation Criteria	Adjustment Strategies*	Constraint Types	Constraint Handling
Reddy, *et al.*, 1999	Genetic Algorithm	(1) Number of machine changes (2) Number of tool changes (3) Number of set-ups (4) Number of constraint violation	(1) Crossover (2) Mutation	(1) Tool approach directions (2) Datum dependency (3) Tolerance (4) Feature interactions (5) Good practice	(1) Test and generation (2) Two-step manipulation for crossover (3) Penalty algorithm
Qiao, *et al.*, 2000	Genetic Algorithm	Part machining time	(1) Crossover (2) Mutation	(1) Fixed order of operations (2) Resource constraints	Tree traversal algorithm
Ma, *et al.*, 2000 Lee, *et al.*, 2001	Simulated Annealing Algorithm	(1) Machine costs (2) Cutting tool costs (3) Number of machine changes (4) Number of tool changes (5) Number of set-ups	(1) Swapping (2) Random exchange (3) Mutation	(1) Fixture constraints (2) Datum dependency (3) Good machining practice	Test and generation

* The adjustment strategies is the next generation strategies for Genetic Algorithm, neighbourhood strategies for Simulated Annealing, or optimised strategies for Neural Networks

Table 5.2 The neighborhood or next-generation strategies in optimisation methods.

Strategies	Descriptions & Examples
Crossover	Two process plans are chosen randomly as parent plans. For them, a cutting point is randomly determined, and each parent plan is separated as left and right parts from the cutting point. A two-step manipulation is applied to generate two children plans that can satisfy the precedence constraints. (i) Copy the left part of parent 1 to the left part of child 1; (ii) In parent 2, find the operations in the right part of parent 1 and copy them to the right part of child 1 according to their sequences in parent 2. Child 2 can be obtained in a similar procedure.

A cutting point

Parent 2 Oper[1] Oper[4] Oper[10] Oper[5] Oper[8] Oper[3] Oper[2] Oper[7] Oper[9] Oper[6]

Parent 1 Oper[9] Oper[2] Oper[4] Oper[1] Oper[10] Oper[7] Oper[8] Oper[6] Oper[5] Oper[3]

Child 1   Oper[9] Oper[2] Oper[4] Oper[1] Oper[10] Oper[5] Oper[8] Oper[3] Oper[7] Oper[6]

Mutation	For a process plan, an operation in the plan is randomly chosen and an alternative operation is used to replace this operation.

Replace | An alternative operation

A plan   Oper[1] Oper[5] Oper[3] Oper[10] Oper[9] Oper[2] Oper[7] Oper[4] Oper[6] Oper[8

Shift	The shift strategy is to remove an operation from its position and insert it at another position in the current plan.

A plan   Oper[1] Oper[5] Oper[3] Oper[10] Oper[9] Oper[2] Oper[7] Oper[4] Oper[6] Oper[8]

Swapping	The swapping strategy is to exchange two operations chosen randomly in a plan.

A plan   Oper[1] Oper[5] Oper[3] Oper[10] Oper[9] Oper[2] Oper[7] Oper[4] Oper[6] Oper[8]

Adjacent Swapping	The adjacent swapping strategy is to exchange two adjacent operations in a plan.

A plan   Oper[1] Oper[5] Oper[3] Oper[10] Oper[9] Oper[2] Oper[7] Oper[4] Oper[6] Oper[8]

Table 5.3 Optimisation methods and their neighborhood or next-generation strategies.

Algorithms	Strategies				
	Crossover	Mutation	Shift	Swapping	Adjacent Swapping
Genetic algorithm	✓	✓			
Simulated annealing		✓	✓	✓	✓
Tabu search		✓	✓	✓	✓

reasoning processes. The test and generation method used by Zhang, *et al.* [1997], Reddy, *et al.* [1999], Ma, *et al.* [2000] and Lee, *et al.* [2001] is to generate process plans randomly, and then test and select some feasible plans for further manipulations. The fundamental problems of this approach include the low efficiency and difficulty to generate reasonable initial plans for a complex part. The two-step manipulation [Zhang, *et al.*, 1997; Reddy, *et al.*, 1999; Qiao, *et al.*, 2000] is specific for the crossover operations, and it is not applicable to SA and TS, or other operations in GA, such as creating initial generations and mutations. The penalty method [Chen, *et al.*, 1998; Reddy, *et al.*, 1999] performed better in terms of computational efficiency and extensible search space. However, the method, which penalises more values to the evaluation function of invalid plans, might cause more time to be spent on evaluating these invalid plans, and the final optimal plan might not be usable. New methods should be studied to handle the possibility of conflicting constraints, which can be classified according to their impact on the feasible manufacturability of process plans, and achieve good computational efficiency and robustness at the same time.

5.2 Knowledge Representation of Process Plans

5.2.1 Process plan representation

A process plan for a part consists of machining operations, applicable alternative machining resources (machines and cutting tools), set-up plans, machining parameters, operation sequence, etc. A set-up is usually defined as a group of operations that are machined on a single machine with the same fixture configuration. Here, a set-up is a group of features with the same TAD machined in the same fixture configuration on a machine. For instance, in Fig. 5.2, a hole with two TADs is considered to be related to two set-ups. A valid TAD should satisfy the following conditions:

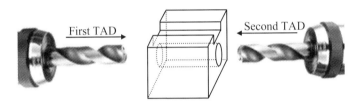

First TAD   Second TAD

Fig. 5.2 A through hole with two TADs.

- Tool accessibility. If a cutting tool for machining a feature on a part along one of its TADs is obstructed by other features on the part, or the cutting tool cannot be positioned to machine the feature along the TAD correctly, the TAD for this feature is considered to be inaccessible and invalid.
- Fixture. If there is no valid fixture element for holding a part to machine a feature along one of its TADs, the feature cannot be fixtured and machined along the TAD, and the TAD is invalid.
- Availability of cutters. Along one of its TADs, if the shape of a feature is beyond the scope of any cutting tools available, the feature cannot be machined along this TAD, and the TAD is invalid.

- Tolerance and surface finish requirements. A feature should not violate the tolerance and surface finish requirements of machines when it is machined along one of its TADs. Otherwise, the TAD is invalid.

The operations and their relevant machines, cutting tools, TADs, and machining parameters can be modelled as a process plan. If there are n operations for machining a part, the plan will be n bits, and each bit represents an operation. The sequence of the bits is the operation sequence of the process plan. An operation has a set of candidate machines, tools, and TADs under which the operation can be executed. In an object-oriented description, a bit (an operation) in a process plan and a process plan can be defined as shown in Tables 5.4 and 5.5.

Table 5.4 Class definitions of an operation (a process plan bit).

Class Process_Plan_Bit (Only containing a variable domain)

Data Types	Variables	Descriptions
int	Operation_id	The id of the operation
int	Machine_id	The id of a machine to execute the operation
int	Tool_id	The id of a cutting tool to execute the operation
int	TAD_id	The id of a TAD to apply the operation
int[ ]	Machine_list[ ]	The candidate machine list for executing the operation
int[ ]	Tool_list[ ]	The candidate tool list for executing the operation
int[ ]	TAD_list[ ]	The candidate TAD list for applying the operation
specific_type	Operation_parameters	Other machining parameters of the operations

Table 5.5 Class definitions of a process plan.

Class Process_Plan (Containing a variable domain and a method domain)

The Variable Domain

Data Types	Variables	Descriptions
Process_Plan_Bit (an operation)	Oper[n]	Define a process plan Oper[n] based on the above class – Process_Plan_Bit. n is the number of operations in the plan.
double	*TMC*	Total Machine Cost of the plan
double	*TTC*	Total Tool Cost of the plan
double	*TSC*	Total Set-up Cost of the plan
double	*TMCC*	Total Machine Change Cost of the plan
double	*TTCC*	Total Tool Change Cost of the plan
double	*APC*	Additional Penalty Cost of violating constraints in the plan
double	*TWC*	Total Weighted Cost of the plan

The Method Domain

Return Types	Methods	Descriptions
double[ ]	Cost_Comp()	The method is used to compute the individual costs and weighted total cost of the plan.
Boolean	Variant_Plan_ Generation()	Two basic manipulations in the method are used to generate variant plans from the plan (neighbourhood strategies).

5.2.2 *Machining cost criteria of process plans*

The total machining cost in a process plan includes all the machine costs, tool costs, machine change costs, set-up costs, tool change costs, and additional penalty costs due to the violation of precedence constraints. These costs can be computed as below.

(1) The Machine Cost (*MC*). *MC* is the total costs of the machines used in a process plan, and it can be computed as:

$$MC = \sum_{i=1}^{n}(Oper[i].Mac\_id * MCI) \tag{5.1}$$

where *MCI* is the machine cost index for a machine.

(2) The Tool Cost (*TC*). *TC* is the total cost of the cutting tools used in a process plan, and it can be computed as:

$$TC = \sum_{i=1}^{n}(Oper[i].Tool\_id * TCI) \tag{5.2}$$

where *TCI* is the tool cost index for a tool.

(3) Number of Set-up Changes (*NSC*), Number of Set-up (*NS*), and Total Set-up Cost (*TSC*). Here, a set-up change between two consecutive operations is defined in Table 5.6, and *NSC* can be computed as:

$$NSC = \sum_{i=1}^{n-1}\Omega_2(\Omega_1(Oper[i].Machine\_id, Oper[i+1].Machine\_id), \tag{5.3}$$
$$\Omega_1(Oper[i].TAD\_id, Oper[i+1].TAD\_id))$$

The corresponding *NS* can be computed as:

$$NS = 1 + NSC \tag{5.4}$$

Table 5.6 The definition of a set-up change.

Conditions of Machining Two Consecutive Operations	A Set-up Change
Same TAD and same machine	No
Same TAD and different machines	Yes
Different TADs and same machine	Yes
Different TADs and different machines	Yes

The $\underline{S}$et-up $\underline{C}$ost (SC) is considered to be the same for each set-up. Hence,

$$TSC = \sum_{i=1}^{NS} SC \qquad (5.5)$$

where $\Omega_1(X,Y) = \begin{cases} 1 & X \neq Y \\ 0 & X = Y \end{cases}$, $\Omega_2(X,Y) = \begin{cases} 0 & X = Y = 0 \\ 1 & otherwise \end{cases}$.

(4) $\underline{N}$umber of $\underline{M}$achine $\underline{C}$hanges (NMC) and $\underline{T}$otal $\underline{M}$achine $\underline{C}$hange $\underline{C}$ost ($TMCC$).

$$NMC = \sum_{i=1}^{n-1} \Omega_1(Oper[i].Machine\_id, Oper[i+1].Machine\_id) \qquad (5.6)$$

The $\underline{M}$achine $\underline{C}$hange $\underline{C}$ost (MCC) is considered to be the same for each machine change. Hence,

$$TMCC = \sum_{i=1}^{NMC} MCC \qquad (5.7)$$

(5) $\underline{N}$umber of $\underline{T}$ool $\underline{C}$hanges (NTC) and $\underline{T}$otal $\underline{T}$ool $\underline{C}$hange $\underline{C}$ost ($TTCC$). A tool change is defined in Table 5.7. NTC is computed as:

$$NTC = \sum_{i=1}^{n-1} \Omega_2(\Omega_1(Oper[i].Machine\_id, Oper[i+1].Machine\_id), \qquad (5.8)$$
$$\Omega_1(Oper[i].Tool\_id, Oper[i+1].Tool\_id))$$

Similarly, the $\underline{T}$ool $\underline{C}$hange $\underline{C}$ost (TCC) is considered to be the same for each tool change. Thus

$$TTCC = \sum_{i=1}^{NTC} TCC \qquad (5.9)$$

(6) $\underline{N}$umber of $\underline{V}$iolating $\underline{C}$onstraints (NVC) and $\underline{A}$dditional $\underline{P}$enalty $\underline{C}$ost (APC).

$$NVC = \sum_{i=1}^{n-1} \sum_{j=i+1}^{n} \Omega_3(Oper[i].Operation\_id, Oper[j].Operation\_id)) \qquad (5.10)$$

A fixed $\underline{P}$enalty $\underline{C}$ost (PC) is applied to each violated constraint. Thus

$$APC = \sum_{i=2}^{NVC} PC \qquad (5.11)$$

where
$$\Omega_3(X,Y) = \begin{cases} 1 & \text{The sequence of } X \text{ and } Y \text{ violates constraints} \\ 0 & \text{The sequence of } X \text{ and } Y \text{ is in accordance to constraints} \end{cases}$$

(7) The Total Weighed Cost (*TWC*).

$$TWC = w_1 * MC + w_2 * TC + w_3 * SC + w_4 * MCC + w_5 * TCC + w6 * APC \quad (5.12)$$

where $w_1 - w_6$ are the weights.

The *MC*, *TC*, *SC*, *MCC*, *TCC* and *PC* are stored in several relational tables of a database system as illustrated in Table 5.8.

Table 5.7 The definition of a tool change.

Conditions Of Machining Two Consecutive Operations	A Tool Change
Same tool and same machine	No
Same tool and different machines	Yes
Different tools and same machine	Yes
Different tools and different machines	Yes

5.2.3 *Precedence constraints*

The geometric and manufacturing interactions between features as well as the technological requirements in a part are considered to generate some preliminary precedence constraints between the operations. Constraints affect the generation of process plans and can be classified as "hard" or "soft" constraints [Faheem, *et al.*, 1998]. Hard constraints affect the manufacturing feasibility and a process plan should be consistent with these constraints. Compared to hard constraints, soft constraints only affect the quality, cost or efficiency of a process plan. These soft constraints can be violated at certain times in cases of contradictions to some hard constraints, to achieve the lowest total machining cost. The classifications, definitions and illustrative examples of precedence constraints are given in Tables 5.9 and 5.10.

Table 5.8 Several tables for storing machining index costs in a database.

Information of machines

Fields	Descriptions
ID	Id of the machine (keyword)
Type	Type of the machine
MC	Machine Cost of the machine

Information of cutting tools

Fields	Descriptions
ID	Id of the cutting tool (keyword)
Type	Type of the tool
TC	Tool Cost of the tool

Cost information of set-ups, machine changes, tool changes and penalties

Fields	Descriptions
SC	Set-up Cost
MCC	Machine Change Cost
TCC	Tool Change Cost
PC	Penalty Cost

The optimisation objective is to select a process plan from the possible candidates to achieve the minimum total weight cost. The optimised process plan should satisfy the condition that there is no violation of hard constraints. Therefore, based on the above concepts, the constrained process planning optimisation model can be represented as follows:

$$Objective: \quad \underset{x}{Min} \quad TWC(x), \text{ and } x \in \{candidate\ process\ plans\}$$

$$s.t. \quad NVC(x) = 0 \text{ for all hard constraints}$$

where, TWC is the total weight cost defined in Equation (5.12), x is a selected process plan from its candidate pool, and NVC is the number of violating constraints defined in Equation (5.10).

In the following content, an intelligent method, namely a hybrid GA/SA method, is elaborated to solve this problem. Case studies and comparisons are given for this method.

Table 5.9 Definitions and classifications of precedence constraints.

Constraints	Definitions
Hard Constraints	
Fixture interactions	The clamping or supporting faces for machining a feature are destroyed by machining another feature earlier.
Tool interactions	The positioning faces required by a cutting tool to machine a feature are removed by the machining of another feature earlier.
Datum interaction	In order to locate a part for machining or inspection, some datum faces in the part are used as reference planes. A datum interaction occurs when machining a feature destroys the datum required for another feature.
Feature priorities	A feature should be machined before its associated features. Another case is that a feature should be machined first to provide entrance face for machining an interacting feature.
Fixed order of machining operations	This case includes some explicit precedence constraints, for example, turning-grooving-chamfering prior to thread cutting.
Soft Constraints	
Thin-wall interactions	A thin-wall interaction occurs when the distance between features is very small and causes precedence constraints in machining.
Material-removal interactions	For two features with geometric interactions, if the different material removal sequences of features influence the cost or the quality of machining and cause precedence constraints between these features, a material-removal interaction occurs.

Table 5.10 Examples of precedence constraints.

Constraints	Examples	Explanations
Hard Constraints		
Fixture interactions	Vise jaw Hole / Chamfer Vise jaw	The hole should be machined before the chamfer, otherwise it cannot be fixtured.
Tool interactions	Chamfer / Hole	In order to position a drilling tool correctly, the drilling of the hole should precede the machining of the chamfer.
Datum interaction	Datum feature (top face) / Machining face	The top face (the datum feature) should be machined prior to the base face.
Feature priorities	Countersunk / Hole / Pocket 1 / Slot / Pocket	(Case 1) The countersunk is an associated feature and should be machined after the primary hole. (Case 2) The two pockets should be machined first to expose the entrance faces of the slot.
Fixed order of machining operations	Operations for a hole: (1) Drilling (2) Boring and (3) Reaming	A typical sequence of machining a hole is drilling-boring and reaming.
Soft Constraints		
Thin-wall interactions	Slot Thin wall / Hole	The good practice should be drilling the hole, then machining the slot to avoid the deformation of the thin wall.
Material-removal interactions	Step / Hole	The step should be machined prior to the hole for achieving high machining efficiency (milling is faster than drilling) and surface quality.

5.3 A Hybrid GA/SA-based Optimisation Method

5.3.1 Overview of the algorithm

It has been recognised that GA is not well-suited for performing the optimisation of individual plans [Rogers, 1991; Ishibuchi, *et al.*, 1994; Mathias, *et al.*, 1994; Renders and Bersini, 1994; Yen, *et al.*, 1998]. Once the good performance regions of a search space have been identified using GA, it may be useful to invoke a local regional search routine to optimise the members of the final population [Grefenstette, 1987]. SA has proven to be effective for finding the optimal or near-optimal solution for a local regional search [Pham and Karaboga, 2000]. To achieve the global optimisation effectively, the strengths of GA and SA have been combined. The overview of the proposed hybrid GA/SA is presented as follows:

(1) Generate the initial GA_k_1 populations and initialise the parameters of the GA.

(2) After m generations computation using the GA, choose GA_k_2 good chromosomes with certain Hamming distances between them as the initial plans for the SA (the current plans in the SA have the same representations as the chromosomes in the GA). The Hamming distances between two chromosomes, for example, $Oper_i[n]$ and $Oper_j[n]$, can be defined as

$$\text{Hamming distance } (HD) = \sum_{l=1}^{n} \Omega_1 (Oper_i[l], Oper_j[l]) \tag{5.13}$$

$$\text{where } \Omega_1(X,Y) = \begin{cases} 1 & X \neq Y \\ 0 & X = Y \end{cases}.$$

(3) Initialise the parameters of the SA.

(4) Apply the SA algorithm to the current plans and compute iteratively until the stopping criterion is met. The updated current plans are the optimal or near-optimal process plans.

A flowchart of the hybrid approach is represented in Fig. 5.3. The details of the approach are described next.

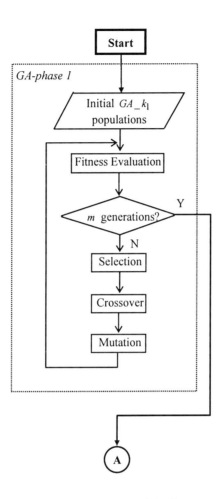

Fig. 5.3 A flowchart of the hybrid GA/SA approach.

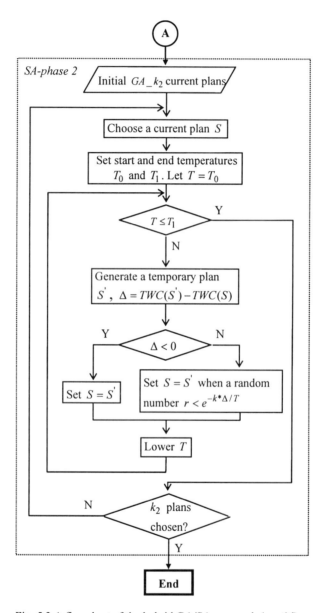

Fig. 5.3 A flowchart of the hybrid GA/SA approach (cont'd).

5.3.2 *Genetic algorithm – phase 1*

(1) Initialisation. Generate GA_k_1 initial populations for a part with n operations. To create each chromosome, a sequence of the n operations is randomly permutated to form a n-bit chromosome. The machine, tool, and TAD in each bit are randomly determined from the corresponding candidate lists. Apply a constraint adjustment algorithm, which will be described later, to the GA_k_1 initial populations to adjust them into the feasible domain.

(2) Fitness evaluation. Since GA is used to achieve the maximum objective, to apply GA to the minimum cost objective in this problem, the fitness function (FF) can be represented as follows:

$$FF = \lambda * (UL - TWC) * (UL - TWC) \tag{5.14}$$

where UL is an upper limit constant for TWC, and λ is a positive coefficient.

(3) Selection. The populations are reproduced for the next generation using some selection strategies. In this algorithm, the "roulette wheel selection" strategy and an "elite" strategy are employed to expedite the search and guarantee the search in a non-decreasing trend.

(4) Crossover. Choose two chromosomes from the GA_k_1 populations as the parent chromosomes for a crossover operation. A cutting point is randomly determined, and each parent chromosome is separated into left and right parts from the cutting point. To ensure that all the operations appear in the child chromosomes once and only once after the crossover, a two-step manipulation [Dagli and Sittisathanchai, 1993; Zhang, *et al.*, 1997; Qiao, *et al.*, 2000] is applied. The two-step manipulation is: (a) copy the left part of Parent 1 to the left part of Child 1, and (b) in Parent 2, find the bits in the right part of Parent 1 and copy them to the right part of child 1 according to their sequence in Parent 2. Child 2 can be obtained similarly. Fig. 5.4 illustrates this procedure for a 10-bit chromosome. Since the child chromosomes follow the precedence constraints between operations in this crossover, the constraint adjustment algorithm is not required. The probability of applying the crossover is defined as P_{GA_c}.

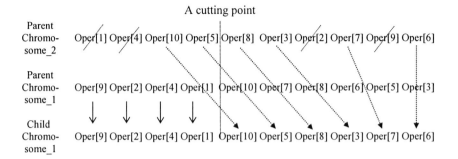

Fig. 5.4 Child chromosome_1 is obtained in a two-step crossover manipulation.

(5) Mutation. Two types of mutation strategies are applied. The first mutation strategy exchanges two operations chosen randomly in a chromosome. The probability of applying this strategy is defined as P_{GA_m1}. After applying this strategy, the constraint adjustment algorithm is applied to the generated populations to adjust them to the feasible domain. The second mutation strategy randomly selects an operation in a chromosome, and replaces the set of machine, tool, and the TAD used in this operation from the candidate lists. The probability of applying this strategy is defined as P_{GA_m2}.

(6) Steps (2) – (5) are repeated after GA_m generations.

5.3.3 *Simulated annealing algorithm – phase 2*

The SA algorithm is the second phase of the hybrid GA/SA approach, and complements the GA in achieving the global optimal or near-optimal process plans by performing a better localised search. Its processes are described as follows:

(1) Choose GA_k_2 chromosomes with good performance and certain Hamming distances between them from the GA generated populations to form the initial current plans for the SA.

(2) Set a chromosome as the current plan S.

(3) Determine the start and end temperatures T_0 and T_1. Set the current temperature T as $T = T_0$.

(4) While not yet frozen ($T > T_1$), perform Steps (a) – (c):

(a) Generate a temporary plan S' by making some random changes using three neighbourhood strategies simultaneously. After these neighbourhood strategies have been applied, apply the constraint handling algorithm to S' to adjust it to the feasible domain. The neighbourhood strategies are:

(i) Shift. This strategy removes an operation from its present position and inserts it at another position in the current plan. The probability of applying this strategy is defined as P_{SA_s}.

(ii) Adjacent swapping. This strategy exchanges two adjacent operations in the current plan. The probability of applying the interchange strategy is defined as P_{SA_as}.

(iii) Mutation. This is the same as that of the GA. The probabilities of applying the first and second mutation strategies are defined as P_{SA_am1} and P_{SA_am2} respectively.

(b) Set

$$\Delta = TWC(S') - TWC(S) \tag{5.15}$$

If $\Delta \leq 0$ (downhill move):

Set $S = S'$.

Else (uphill move):

Choose a random number r from [0, 1]. Set $S = S'$ when

$$r < e^{-k^*\Delta/T} \tag{5.16}$$

(5) Return to step (4) after lowering the temperature T as

$$T = \alpha * T \tag{5.17}$$

where $0 < \alpha < 1$.

(6) If frozen ($T \leq T_1$), return to Step (2) until all SA_k_2 plans have been computed.

The cooling schedule used in Equation (5.17) is taken from [Dowsland, 1995; Cho, *et al*, 1998]. In practice, there are several cooling schedules that are widely used [Pham and Karaboga, 1998]. In order to achieve optimal or near-optimal solutions, the cooling process has to be very slow, and α in Equation (5.17) should be very close to 1.

5.3.4 *Constraint handling algorithm*

For an initially generated or an adjusted process plan (a chromosome in the GA or a current/temporary plan in the SA), after the crossover, mutation, or neighbourhood strategies in the GA or SA have been performed, the precedence constraints might not be satisfied [Michalewicz and Janikow, 1991]. A two-step hybrid method has been developed to handle hard and soft constraints.

[*STEP 1*] An adjustment algorithm is imposed on a plan (an initial plan or variant plan generated by neighbourhood strategies) to ensure its consistency with the hard constraints.

[*STEP 2*] After *STEP 1*, the optimisation model is transformed from a constrained problem to an unconstrained problem where the weighted sum of the various machining costs and the additional penalty costs caused by the soft constraints is minimised. Under this scheme, soft constraints can be violated and compromised to achieve the lowest total machining cost from the multiple criteria.

The constraint adjustment algorithm in *STEP 1*, which can be applied to a complicated and multiple constraint condition, is proposed to re-arrange the bits in a process plan according to the hard constraints. For a process plan PP with n bits (operations), the constraint adjustment algorithm is as follows and its flowchart is shown in Fig. 5.5.

(1) Select the bits that do not have constraint relationships with other bits in PP and keep their positions unchanged in PP. Let the number of bits selected in this step be n_1.

(2) The remaining ($n - n\_1$) bits, which are constrained to be prior to or posterior to other bits in PP, are used to form a double-linked list (LL) according to their relative positions in PP. The adoption of a double-linked list is to make deletion and insertion manipulations convenient and efficient. In LL, each bit has prior and next references, pointing to its prior and next bits respectively.

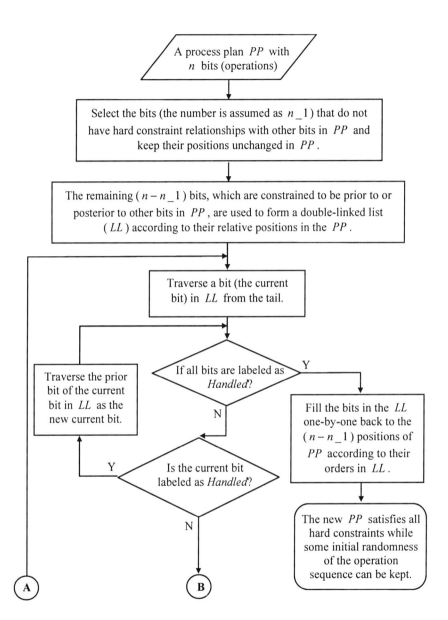

Fig. 5.5 The workflow of a constraint adjustment algorithm for hard constraints.

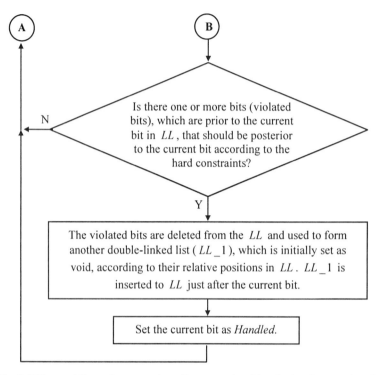

Fig. 5.5 The workflow of a constraint adjustment algorithm for hard constraints (cont'd).

(3) Traverse *LL* from the tail. Set the traversed node as *the current bit* if it has not been assigned as *Handled*, otherwise *the current bit* is moved to the bit that it has prior reference to. If there is one or more bits, which are prior to *the current bit* in the *LL*, they should be posterior to *the current bit* according to the preliminary precedence constraints. These bits are deleted from the *LL* and used to form another double-linked list (*LL* _1), which is initially set as void, according to their relative positions in *LL*. *LL* _1 is inserted to *LL* just after *the current bit*. Move the reference to the tail, and set the just handled *current bit* as *Handled*. Repeat this step.

(4) After all the bits have been assigned as *Handled* in Step (3), the order in *LL* reflects the proper relative precedence relationships of the constrained bits.

(5) Fill the bits in the *LL* one-by-one back into the ($n - n\_1$) positions of *PP* according to their orders in *LL* . The updated *PP* satisfies the precedence constraints while some randomness can be kept.

For example, for a 14-bit chromosome ($n = 14$), the bits sequence and precedence hard constraints are listed in Table 5.11. Six bits, namely, Oper[1], Oper[4], Oper[6], Oper[11], Oper[13] and Oper[14], have no constraint relationships with other operations ($n\_1 = 6$). Hence, their positions are kept and a *LL* is formed for the other eight bits ($n - n\_1 = 8$). The first *current bit* is Oper[8], and Oper[3], Oper[9], Oper[5] should be posterior to it according to the constraints. The updating process of *LL* is illustrated in Fig. 5.6. After Oper[8] has been handled, the reference to *the current bit* is moved to the tail and the same procedure is continued until all bits are assigned as *Handled*. The finally updated process plan satisfies all the hard constraints.

5.4 Experimental Results

5.4.1 *Sample parts*

Two prismatic parts are used for the case studies. The first prismatic part (Part 1) used by Zhang *et al.* [1997] is illustrated in Fig. 5.7. It consists of 14 STEP-defined manufacturing features and machining operations ($n = 14$). The relevant information for the machining resources, features and operations, and the precedence constraints are given in Tables 5.12-5.14.

Table 5.11 A process plan with four hard constraints.

Original process plan	Oper[7]-Oper[14]-Oper[2]-Oper[10]-Oper[4]-Oper[11]-Oper[9]-Oper[12]-Oper[3]-Oper[13]-Oper[6]-Oper[5]-Oper[8]-Oper[1]
Constraint 1	Oper[5] and Oper[9] should be prior to Oper[2] and Oper[7]
Constraint 2	Oper[12] and Oper[8] should be prior to Oper[3], Oper[5] and Oper[9]
Constraint 3	Oper[3] should be prior to Oper[5]
Constraint 4	Oper[10] should be prior to Oper[7]

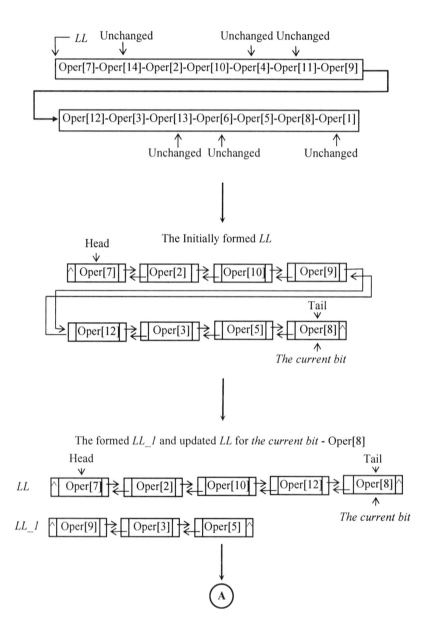

Fig. 5.6 An example process of the constraint adjustment algorithm.

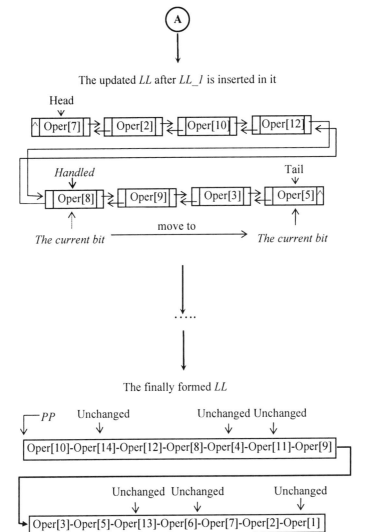

Fig. 5.6 An example process of the constraint adjustment algorithm (cont'd).

Table 5.12 Available machining resources and costs in a workshop environment for Part 1.

Machines

No.	Types	MC
M_1	Drill press	10
M_2	Milling machine	35
M_3	Three-axis vertical milling machine	60

Tools

No.	Types	TC
C_1	Drill2	3
C_2	Drill12	3
C_3	Reamer	8
C_4	Boring tool	15
C_5	Milling cutter 1	10
C_6	Milling cutter 2	15
C_7	Slot cutter	10
C_8	Chamfer tool	10

$MCC = 300, SC = 120, TCC = 15, PC = 100$

The second prismatic part (Part 2) is from [Shah, *et al.*, 1995] and illustrated in Fig. 5.8. It has more complex features and constraints and consists of 14 STEP-defined manufacturing features and 20 machining operations ($n = 20$). The relevant information for machining resources, features and operations, and precedence constraints are given in Tables 5.15-5.17.

Table 5.13 The features, operations and candidate machining information for Part 1.

Features	Feature Descriptions	Operations (Oper_id)	TAD Candidates	Machine Candidates	Tool Candidates
F_1	Two holes as a replicated feature	Drilling (Oper$_1$)	+z, -z	M_1, M_2, M_3	C_1
F_2	A chamfer	Milling (Oper$_2$)	-x, +y, -y, -z	M_2, M_3	C_8
F_3	A slot	Milling (Oper$_3$)	+y	M_2, M_3	C_5, C_6
F_4	A slot	Milling (Oper$_4$)	+y	M_2	C_5, C_6
F_5	A step	Milling (Oper$_5$)	+y, -z	M_2, M_3	C_5, C_6
F_6	Two holes as a replicated feature	Drilling (Oper$_6$)	+z, -z	M_1, M_2, M_3	C_2
F_7	Four holes as a replicated feature	Drilling (Oper$_7$)	+z, -z	M_1, M_2, M_3	C_1
F_8	A slot	Milling (Oper$_8$)	+x	M_2, M_3	C_5, C_6
F_9	Two holes as a replicated feature	Drilling (Oper$_9$)	-z	M_1, M_2, M_3	C_1
F_{10}	A slot	Milling (Oper$_{10}$)	-y	M_2, M_3	C_5, C_6
F_{11}	A slot	Milling (Oper$_{11}$)	-y	M_2, M_3	C_5, C_7
F_{12}	Two holes as a replicate feature	Drilling (Oper$_{12}$)	+z, -z	M_1, M_2, M_3	C_1
F_{13}	A step	Milling (Oper$_{13}$)	-x, -y	M_2, M_3	C_5, C_6
F_{14}	Two holes as replicate feature	Drilling (Oper$_{14}$)	-y	M_1, M_2, M_3	C_1

Table 5.14 The precedence constraints for Part 1.

Constraints	Descriptions	Hard or Soft
Tool interactions	$Oper_1$ should be prior to $Oper_2$	Hard
Datum interactions	$Oper_6$ should be prior to $Oper_7$. $Oper_{10}$ should be prior to $Oper_{11}$. $Oper_{13}$ should be prior to $Oper_{14}$.	Hard
Thin-wall interactions	$Oper_9$ should be prior to $Oper_8$. $Oper_{12}$ should be prior to $Oper_{10}$.	Soft
Material removal interactions	$Oper_8$ should be prior to $Oper_9$. $Oper_{10}$ should be prior to $Oper_{12}$. $Oper_{13}$ should be prior to $Oper_{14}$. $Oper_3$ should be prior to $Oper_4$.	Soft

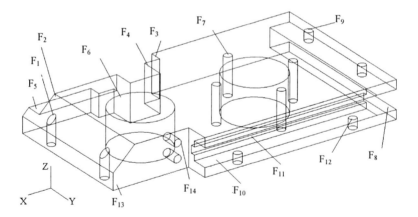

Fig. 5.7 A sample part with 14 features – Part 1.

Table 5.15 The information for available machines and cutting tools for Part 2.

Machines

No.	Types	*MC*
M_1	Drilling press	10
M_2	3-axis vertical milling machine	40
M_3	CNC 3-axis vertical milling machine	100
M_4	Boring machine	60

Cutting Tools

No.	Types	*TC*
C_1	Drill 1	7
C_2	Drill 2	5
C_3	Drill 3	3
C_4	Drill 4	8
C_5	Tapping tool	7
C_6	Mill 1	10
C_7	Mill 2	15
C_8	Mill 3	30
C_9	Ream	15
C_{10}	Boring tool	20

$MCC = 160$, $SC = 100$, $TCC = 20$, $PC = 100$

Table 5.16 The features and operations information for Part 2.

Feat-ures	Feature Descriptions	Operations (Oper_id)	TAD Candidates	Machine Candidates	Tool Candidates
F_1	A planar surface	Milling (Oper$_1$)	+z	M_2, M_3	C_6, C_7, C_8
F_2	A planar surface	Milling (Oper$_2$)	-z	M_2, M_3	C_6, C_7, C_8
F_3	Two pockets arranged as a replicated feature	Milling (Oper$_3$)	+x	M_2, M_3	C_6, C_7, C_8
F_4	Four holes arranged as a replicated feature	Drilling (Oper$_4$)	+z, -z	M_1, M_2, M_3	C_2
F_5	A step	Milling (Oper$_5$)	+x, -z	M_2, M_3	C_6, C_7
F_6	A protrusion (rib)	Milling (Oper$_6$)	+y, -z	M_2, M_3	C_7, C_8
F_7	A boss	Milling (Oper$_7$)	-a	M_2, M_3	C_7, C_8
F_8	A compound hole	Drilling (Oper$_8$) Reaming (Oper$_9$) Boring (Oper$_{10}$)	-a	M_1, M_2, M_3 M_1, M_2, M_3 M_3, M_4	C_2, C_3, C_4 C_9 C_{10}
F_9	A protrusion (rib)	Milling (Oper$_{11}$)	-y, -z	M_2, M_3	C_7, C_8
F_{10}	A compound hole	Drilling (Oper$_{12}$) Reaming (Oper$_{13}$) Boring (Oper$_{14}$)	-z	M_1, M_2, M_3 M_1, M_2, M_3 M_3, M_4	C_2, C_3, C_4 C_9 C_{10}
F_{11}	Nine holes arranged in a replicated feature	Drilling (Oper$_{15}$) Tapping (Oper$_{16}$)	-z	M_1, M_2, M_3 M_1, M_2, M_3	C_1 C_5
F_{12}	A pocket	Milling (Oper$_{17}$)	-x	M_2, M_3	C_7, C_8
F_{13}	A step	Milling (Oper$_{18}$)	-x, -z	M_2, M_3	C_6, C_7
F_{14}	A compound hole	Reaming (Oper$_{19}$) Boring (Oper$_{20}$)	+z	M_1, M_2, M_3 M_3, M_4	C_9 C_{10}

Table 5.17 Precedence constraints between machining operations for Part 2.

Features	Operations (Oper_id)	Precedence Constraint Descriptions	Constraints
F_1	Milling (Oper$_1$)	F_1 (Oper$_1$) is the datum and supporting face for the part, hence it is machined prior to all features and operations.	Hard
F_2	Milling (Oper$_2$)	F_2 (Oper$_2$) is prior to F_{10} (Oper$_{12}$, Oper$_{13}$, Oper$_{14}$) and F_{11} (Oper$_{15}$ Oper$_{16}$) for the material removal interactions.	Soft
F_3	Milling (Oper$_3$)		
F_4	Drilling (Oper$_4$)		
F_5	Milling (Oper$_5$)	F_5 (Oper$_5$) is prior to F_4 (Oper$_4$) and F_7(Oper$_7$) for the datum interactions.	Hard
F_6	Milling (Oper$_6$)	F_6 (Oper$_6$) is prior to F_{10} (Oper$_{12}$, Oper$_{13}$, Oper$_{14}$) for the datum interaction.	Hard
F_7	Milling (Oper$_7$)	F_7 (Oper$_7$) is prior to F_8 (Oper$_8$, Oper$_9$, Oper$_{10}$) for the datum interactions.	Hard
F_8	Drilling (Oper$_8$) Reaming (Oper$_9$) Boring (Oper$_{10}$)	Oper$_8$ is prior to Oper$_9$ and Oper$_{10}$, Oper$_9$ is prior to Oper$_{10}$ for the fixed order of machining operations.	Hard
F_9	Milling (Oper$_{11}$)	F_9 (Oper$_{11}$) is prior to F_{10} (Oper$_{12}$, Oper$_{13}$, Oper$_{14}$) for the datum interaction.	Hard
F_{10}	Drilling (Oper$_{12}$) Reaming (Oper$_{13}$) Boring (Oper$_{14}$)	Oper$_{12}$ is prior to Oper$_{13}$ and Oper$_{14}$, Oper$_{13}$ is prior to Oper$_{14}$ for the fixed order of machining operations.	Hard
		F_{10} (Oper$_{12}$, Oper$_{13}$, Oper$_{14}$) is prior to F_{11} (Oper$_{15}$, Oper$_{16}$), and Oper$_{12}$ of F_{10} is prior to F_{14} (Oper$_{19}$, Oper$_{20}$) for the datum interaction.	Hard
F_{11}	Drilling (Oper$_{15}$) Tapping (Oper$_{16}$)	Oper$_{15}$ is prior to Oper$_{16}$ for the fixed order of operations.	Hard
F_{12}	Milling (Oper$_{17}$)		
F_{13}	Milling (Oper$_{18}$)	F_{13} (Oper$_{18}$) is prior to F_4 (Oper$_4$) and F_{12} (Oper$_{17}$) for the material removal interaction.	Soft
F_{14}	Reaming (Oper$_{19}$) Boring (Oper$_{20}$)	Oper$_{19}$ is prior to Oper$_{20}$ for the fixed order of machining operations.	Hard

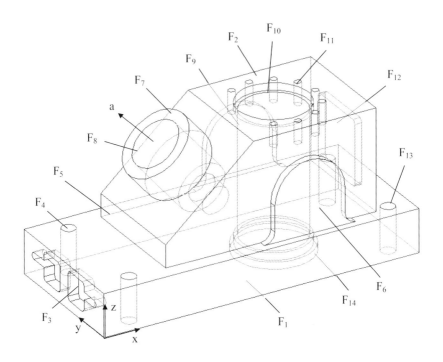

Figure 5.8 A sample part with 14 features – Part 2.

5.4.2 *Computation results*

5.4.2.1 Determination of parameters

For the GA, the main parameters to be determined include the number of populations GA_k_1, the number of generations GA_m, the crossover probability P_{GA_c}, the mutation probabilities P_{GA_m1} and P_{GA_m2}, and λ in Equation (5.14).

If GA_k_1 is too large, the computation time for each iteration will be very long. On the other hand, if GA_k_1 is too small, the optimisation rate will become very slow and not enough number of chromosomes with certain *HD* between them can be generated as the initial SA current plans.

From Fig. 5.9, the number of populations GA_k_1 and generations m were determined as 150 and 100 respectively. When $P_{GA\_c} = 0.65$, $P_{GA\_m1} = 0.1$ and $P_{GA\_m2} = 0.65$, the algorithm can achieve a better performance. Different values of λ have no distinct influence on the final results. Here, $\lambda = 5000.0$.

The main parameters in the SA include the start and end temperatures T_0 and T_1, the Hamming distance *HD* in Equation (5.13), k in Equation (5.16), the annealing rate coefficient α in Equation (5.17), the neighbourhood strategy probabilities P_{SA_s}, P_{SA_as}, P_{SA_am1} and P_{SA_am2}, and the initial number of current plans - SA_k_2.

If *HD* is too large, not enough good initial process plans can be generated for the SA. If it is too small, the initial plans will be too similar. Here, *HD* is determined as 3. Since the annealing process is required to be very slow, optimal or near-optimal solutions can be achieved, $\alpha = 0.9995$. From Fig. 5.10, it can be observed that when $T_0 \approx Highest\_TWC / 10.0$, $k = 10$, and $T_1 \approx T_0 * \alpha^{1000}$, the curves can achieve a better performance or reach the optimum solution. Trials have determined $P_{SA\_as} = P_{SA\_am2} = 0.30$, $P_{SA\_am1} = 0.25$ and $P_{SA\_s} = 0.35$. The initial number of current plans GA_k_2 is chosen as 10.

The parameters have been tested with several other cases to ensure that they are applicable in other situations. For example, in this case study, when the generation of the GA $m \approx 80$, the GA has already converged and should be converted to the SA. However, m was finally determined as 100 in order to meet the requirements of different cases. Other parameters have been handled similarly. In the following Section 5.4.2.2 results are made for Part 2 for demonstration of the method.

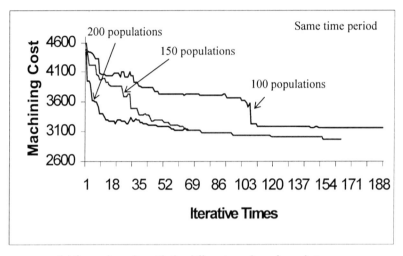

(a) *Several results with the different number of populations*

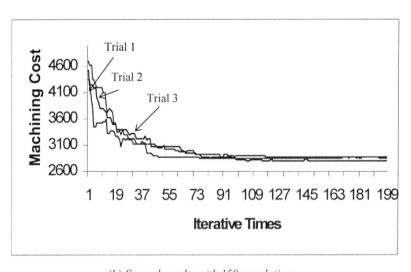

(b) *Several results with 150 populations*

Fig. 5.9 The determination of parameters in GA.

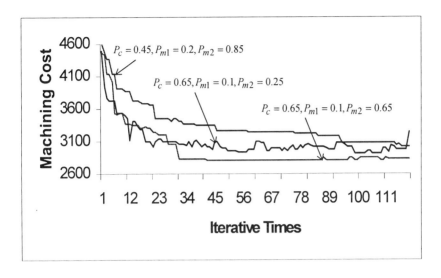

(c) *Several results with the different* P_c, P_{m1}, P_{m2}

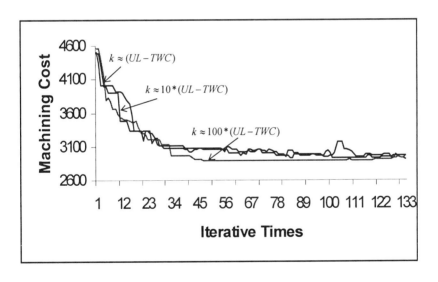

(d) *Several results with the different k*

Fig. 5.9 The determination of parameters in GA (cont'd).

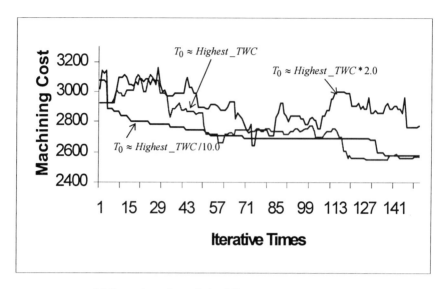

(a) *Several results with the different starting temperatures*

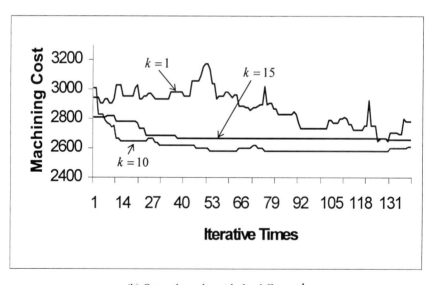

(b) *Several results with the different* k

Fig. 5.10 The determination of parameters in SA.

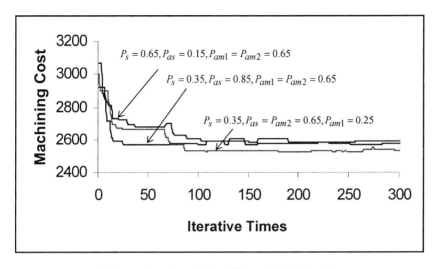

(c) *Several results with the different* $P_s, P_{as}, P_{am1}, P_{am2}$

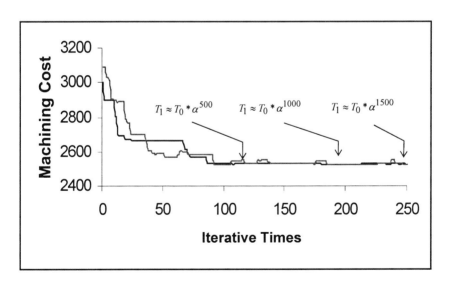

(d) *Several results with the different* T_1

Figure 5.10 The determination of parameters in SA (cont'd).

5.4.2.2 Results under different conditions and criteria

(1) All machines and tools are available, and $w_1 - w_5$ in Equation (5.12) are set as 1.

The results of this approach are shown in Fig. 5.11 (in order to show the curve clearly, only four SA curves are illustrated). Some results are listed in Table 5.18. The best process plan (lowest machining cost) can be generated while the other plans are also satisfactory.

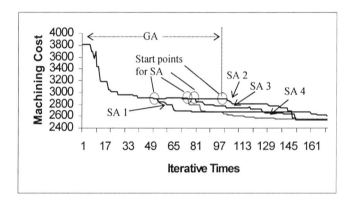

Fig. 5.11 Results of case (1): all machines and tools are available, and $w_1 - w_5 = 1$.

(2) All machines and tools are available, and $w_1 = w_3 = w_4 = 1$, $w_2 = w_5 = 0$.

The machine costs, the numbers of set-ups and the number of machine changes mainly contribute to the total machining cost. Some results are listed in Table 5.19 when only these three costs are considered.

(3) Machine M_2 and Tool C_7 are down, $w_2 = w_5 = 0, w_1 = w_3 = w_4 = 1$.

In a dynamic workshop environment, some machines or tools may be in the state of bottleneck usage or breakdown. Some results are listed in Table 5.20 when M_2 and C_7 are down and certain aspects of the machining costs are considered. All the ten generated process plans achieve the lowest or near-lowest machining cost.

5.4.2.3 *Comparisons with single GA and SA approaches*

With all the machines and tools available, and $w_1 - w_5$ in Equation (5.12) are set as 1, the results from the hybrid approach, and the separated single GA and single SA approaches are shown in Fig. 5.12 for comparison. It can be observed that it is not easy to obtain the optimal results using the single GA approach, while the hybrid approach and the SA approach can reach the optimal or near-optimal results.

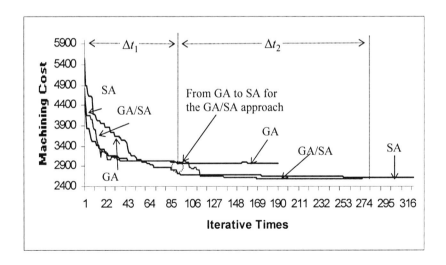

Fig. 5.12 Comparisons of the hybrid GA-SA approach, single GA approach, and single SA approach.

Table 5.18 Some generated process plans of Condition (1).

Process plan 1

Operation	1	3	5	6	11	2	12	13	18	17	7	8	9	19	10	14	20	4	15	16
Machine	2	2	2	2	2	2	2	2	2	2	2	2	2	2	4	4	4	1	1	1
Tool	7	7	7	7	7	2	9	7	7	3	9	9	9	9	10	10	10	2	1	5
TAD	+z	+x	+x	-z	-z	-z	-z	-x	-x	-a	-a	-a	-a	+z	-a	-z	+z	-z	-z	-z

$NMC = 2$, $NS = 10$, $NTC = 9$, $MCC = 320$, $SC = 1000$, $TCC = 180$, $MC = 770$, $TC = 267$

Total Cost: 2537.0

Process plan 2

Operation	1	3	5	6	2	11	12	13	18	17	7	8	9	19	10	14	20	15	16	4
Machine	2	2	2	2	2	2	2	2	2	2	2	2	2	2	4	4	4	1	1	1
Tool	7	7	7	7	7	7	2	9	7	7	3	9	9	9	10	10	10	1	5	2
TAD	+z	+x	+x	-z	-z	-z	-z	-x	-x	-a	-a	-a	-a	+z	-a	-z	+z	-z	-z	-z

$NMC = 2$, $NS = 10$, $NTC = 9$ $MCC = 320$, $SC = 1000$, $TCC = 180$, $MC = 770$, $TC = 265$

Total Cost: 2535.0

Table 5.18 Some generated process plans of Conditions (1) (cont'd).

Process plan 3

Operation	1	3	5	6	2	18	11	12	13	17	7	8	9	19	14	20	10	4	15	16
Machine	2	2	2	2	2	2	2	2	2	2	2	2	2	2	4	4	4	1	1	1
Tool	6	6	6	6	6	6	7	3	9	7	7	9	9	10	10	10	10	2	1	5
TAD	+z	+x	-z	-z	-z	-z	-z	-z	-z	-x	-a	-a	-a	+z	-z	+z	-a	-z	-z	-z

$NMC = 2, NS = 10, NTC = 10$ $MCC = 320, SC = 1000, TCC = 200, MC = 770, TC = 237$

Total Cost: 2527.0

Process plan 4

Operation.	1	3	5	6	2	18	11	4	12	13	17	7	8	9	19	10	14	20	15	16
Machine	2	2	2	2	2	2	2	2	2	2	2	2	2	2	2	4	4	4	1	1
Tool	6	6	6	6	6	6	7	2	9	8	8	8	3	9	9	10	10	10	1	5
TAD	+z	+x	-z	-z	-z	-z	-z	-z	-z	-x	-x	-a	-a	-a	+z	-a	+z	+z	-z	-z

$NMC = 2, NS = 10, NTC = 9$ $MCC = 320, SC = 1000, TCC = 180, MC = 800, TC = 267$

Total Cost: 2567.0

In the 10 final plans, the minimum cost is 2527.0 (the best plan), the maximum cost is 2585.0, the average cost is 2546.0.

Table 5.19 Some generated process plans of Condition (2).

Process plan 1

Operation	1	5	11	18	4	6	2	12	13	17	7	8	9	3	10	14	15	16	19	20
Machine	2	2	2	2	2	2	2	2	2	2	2	2	2	2	3	3	3	3	3	3
Tool	7	7	8	7	2	7	6	4	9	7	7	4	9	6	10	10	1	5	9	10
TAD	+z	-z	-z	-z	-z	-z	-z	-z	-z	-x	-a	-a	-a	+x	-a	-z	-z	-z	+z	+z

$NMC = 1, NS = 8$ $MCC = 160, SC = 800, MC = 1160$ Total Cost: 2120.0

Process plan 2

Operation	1	18	17	5	3	4	6	2	11	12	13	7	8	9	10	14	15	16	19	20
Machine	2	2	2	2	2	2	2	2	2	2	2	2	2	2	3	3	3	3	3	3
Tool	7	7	6	6	6	2	7	7	8	4	9	8	8	3	10	10	1	5	9	10
TAD	+z	-x	+x	+x	+x	-z	-z	-z	-z	-z	-z	-a	-a	-a	-a	-z	-z	-z	+z	+z

$NMC = 1, NS = 8$ $MCC = 160, SC = 800, MC = 1160$ Total Cost: 2120.0

The achieved minimum cost: 2120.0.

Table 5.20 Some generated process plans of Condition (3).

Process plan 1

Operation	1	6	18	2	11	12	13	5	14	3	17	7	8	9	10	19	20	15	16	4
Machine	3	3	3	3	3	3	3	3	3	3	3	3	3	3	3	3	3	1	1	1
Tool	6	6	6	8	8	3	9	6	10	8	8	8	4	9	10	9	10	1	5	2
TAD	-z	-z	-z	-z	-z	-z	-z	-z	-z	+x	-x	-a	-a	-a	-a	+z	+z	-z	-z	-z

NMC = 1, NS = 7   MCC = 160, SC = 700, MC = 1730

Total Cost: 2590.0

Process plan 2

Operation	1	11	6	5	2	12	13	14	15	16	18	4	17	3	7	8	9	10	19	20
Machine	3	3	3	3	3	3	3	3	3	3	3	3	3	3	3	3	3	3	3	3
Tool	8	8	6	6	6	4	9	10	5	6	6	2	8	8	8	8	3	8	9	10
TAD	+z	-z	-z	-z	-z	-z	-z	-z	-z	-z	-z	-z	-z	-z	-x	-a	+x	-a	-a	+z

NMC = 0, NS = 6   MCC = 0, SC = 600, MC = 2000

Total Cost: 2600.0

The 10 final plans all achieve the minimum or near-minimum costs: 2590.0 and 2600.0

The iterative rates to obtain an optimal or near-optimal process plan for the hybrid approach and the single SA approach are almost the same, but the hybrid approach can obtain better results (27 of 30 trials reached below 2570.0) than the SA approach (21 of 30 trials reached below 2570.0). Another advantage of the hybrid approach is that it is more efficient than the SA approach in achieving multiple process plans. For instance, in Fig. 5.12, the time for using the hybrid approach to obtain 10 process plans is $\Delta t_1 + 10 * \Delta t_2$, while the time using the SA approach for 10 plans is $10 * (\Delta t_1 + \Delta t_2)$.

5.4.3 *Comparisons of constraint handling methods*

The graph-based heuristic algorithm [Vancza and Markus, 1991], the tree traversal algorithm [Yip-Hoi and Dutta 1996], and the test and generation method [Zhang, *et al.*, 1997; Reddy, *et al.*, 1999; Ma, *et al.*, 2000; Lee, *et al.*, 2001] cannot work well on Part 1, which has conflicting constraints, and Part 2, which has with complex constraints. The proposed hybrid constraint handling method in this research and the penalty method [Chen, *et al.*, 1998; Reddy, *et al.*, 1999], which are applicable to complex situations with conflicting constraints, are used for comparing the computation performance. The results are shown in Fig. 5.13. For the two parts, it can be concluded that the proposed constraint handling method ensures that the computational process is conducted in a smoother and more efficient way.

5.4.4 *Algorithm implementation*

The intelligent method has been implemented using Java 1.4 under a JDK environment. A prismatic part can be created using Unigraphics V15.0 and input to a feature recognition prototype system developed using C++ to extract the manufacturing features and the relevant information of a part. Information of machining resources and machining operations for the generated features are stored in an MS ACCESS database. Through JDBC and SQL, the related machining and operation information from the database can be retrieved and provided for the process planning optimisation module, which can support three alternative methods, namely, the GA, SA and hybrid GA-SA approaches, for generating optimised process plans. A sketch of the information flow is shown in Fig. 5.14.

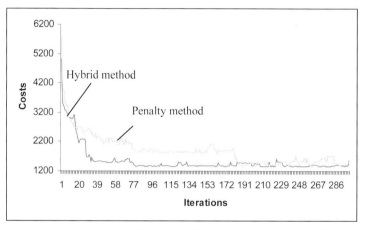

(a) *Machining costs of the intermediate current plans for Part 1.*

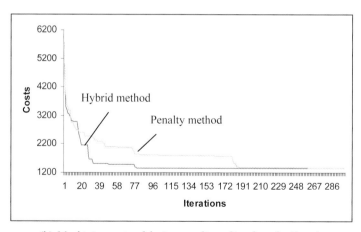

(b) *Machining costs of the intermediate elite plans for Part 1.*

Fig. 5.13. Comparison studies of two constraint handling methods for two sample parts.

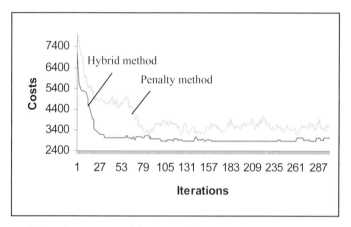

(c) *Machining costs of the intermediate current plans for Part 2*

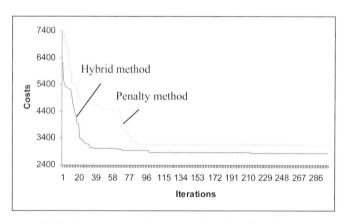

(d) *Machining costs of the intermediate elite plans for Part 2*

Fig. 5.13 Comparison studies of two constraint handling methods for two sample parts (cont'd).

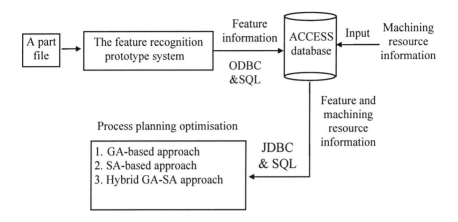

Fig. 5.14 Information flow in the system.

5.5 Summary

A CAPP system, which plays a crucial role to link CAD and CAM processes effectively, has to automatically determine and optimise the process plans for the manufacturing features extracted from a design part, so as to machine the part economically and competitively as well as achieve the desired functional specifications. One of the major difficulties for a generative CAPP system is the selection of suitable set-up plans and machining resources, and sequencing the machining operations so that the least machining cost of the part can be obtained. On the other hand, for a prismatic part, the geometric and manufacturing relationships between manufacturing features are usually complex, which can cause complex constraints in optimising planning. Therefore, the optimisation of the process plans for prismatic parts involves some complex and contradicting decision-making processes. To achieve a global optimal process plan, it is necessary to simultaneously carry out the different decision processes according to a certain evaluation criterion. An optimisation strategy is needed to realise this intractable task successfully. In this chapter, an intelligent method, namely, a hybrid

GA/SA-based method, has been developed to solve this intractable optimisation problem of process planning for prismatic parts by concurrently considering the assignment of machining resources, selection of set-up plans, and sequencing of machining operations.

The advantages of the method can be summarised as the following three points:

(1) The proposed method can generate multiple optimal or near-optimal process plans for a prismatic part with good computation efficiency based on a combined machining cost criterion with weights. Based on the multiple process plans, process planners can make a more accurate and flexible decision according to the actual conditions.

(2) According to the affects on the plan feasibility, precedence constraints are defined and classified. The developed constraint handling method can address a complicated situation with conflicting constraints and achieve good computation performance. The operations are always conducted in a feasible solution domain so that the risk of generating invalid process plans is avoided.

(3) The approaches can conveniently simulate a practical dynamic workshop environment, considering the unavailability of a machine or tool in bottleneck (competition) usage or breakdown, change of machining cost evaluation strategy, and substitution of machines or tools in another shop floor.

Chapter 6

Collaborative Computer-Aided Design – State-of-the-Art

Presently, research is actively being carried out to develop methodologies and technologies for collaborative product development systems to support design teams, which are geographically dispersed, based on the quickly evolving Information Technologies (IT) to facilitate rapid product realisation processes. Many research and commercial systems have been reported to provide collaborative and distributed solutions from the perspectives of CAD, PDM (Product Data Management), workflow management, management of portfolio, engineering and geometric data streaming communication, distributed system infrastructure, PLM (Product Lifecycle Management), etc., and the practical applications are becoming more pervasive and mature.

In this chapter, the related works on collaborative CAD and product design systems are reviewed from three aspects, namely, visualisation-based collaborative systems, co-design collaborative systems (these two form horizontal design collaboration) and hierarchical collaborative systems. From these aspects, the current research and development statuses and research issues, underlying major algorithms, mechanisms and system architectures, and the future trends and challenges are discussed in detail.

6.1 Introduction

During the past two decades, the mechanical CAD industry has experienced some major technological innovations and paradigm shifts. Nowadays, Internet-based collaboration is a popular buzzword in engineering. Fuelled by the quickly evolving IT, the most recent research and development starting from the end of the last century is to renovate CAD and design systems to be distributed and collaborative. It aims to meet the increasing demands of globally joint design and the outsourcing trends in manufacturing. In a collaborative system, designers and engineers can share their works with globally distributed colleagues via the Internet/Intranet. Furthermore, these collaborative systems allow designers to work closely with suppliers, manufacturing partners and customers across the enterprises' firewalls to obtain valuable input into the design chain. With a broader vision, collaborative CAD systems, CAE (Computer-Aided Engineering), CAM, ERP (Enterprise Resource Planning) and PDM systems can be integrated to form collaborative product commerce, which supports intra- and inter-enterprise applications for the whole product life-cycle, as shown in Fig. 6.1. These features provide the ingredients of a product development process to be in accord with the mantra of "cheaper, faster and better".

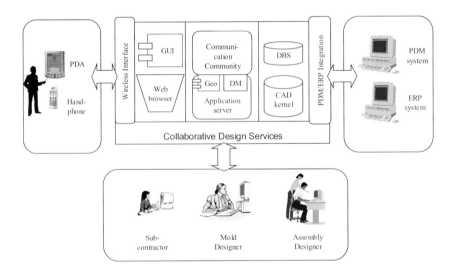

Fig. 6.1 The scenario of collaborative product development.

Recently, research has been actively conducted to develop prototype systems and methodologies for collaborative product development. At the same time, CAD and software vendors have realised the huge business opportunities in this area and there is an intense competition to launch collaborative and distributed CAD software packages. Many big CAD vendors, such as PTC, UGS and Autodesk, are already providing a variety of such products. Software technology companies, such as CoCreate and RealityWorks, have created their own versions of collaborative products. These packages differ in pricing, data transferring methods, collaborative strategies and tools provided, various CAD applications and file formats supported, etc. Several products are briefly discussed next with respects to the ways they offer and create values to the spectrum of design processes.

UGS Inc.'s collaborative product, E-vis™ (www.evis.com), provides visualisation-based collaboration capabilities to users via Web browsers. It has rich real-time information sharing tools, as well as a secure Web-enabled real-time workspace that allows design teams to access multiple data repositories for product data management and visual vaults. Its collaborative viewing capabilities allow multiple users to visualise and share digital design models and mock-ups. In addition, it provides EDS and Microsoft solutions to deliver a flexible combination of Web services to support a wider distributed design chain.

ConceptStation™ (www.realitywave.com), one of many collaboration products developed by RealityWave Inc, focuses on providing an application data transmission technique, namely interactive 3D streaming, to enhance collaborative product development. The streaming technique allows effective and efficient dispatch and access of large-volume industrial data as a series of patched streams, regardless of the user or data location. Powered by this technique, ConceptStation™ supports advanced 2D and 3D viewers with the enhanced visualisation effect and mark-up function for design models. At the same time, it has interfaces to PDM for data retrieval and association over the Internet, and real-time messaging capabilities. Fig. 6.2 illustrates a consecutive "streaming transmission" process of a complex design model.

CoCreate Inc.'s OneSpace™ (www.onespace.net) provides three functions to organise a collaborative product design activity, i.e., an on-line project team workspace, a meeting centre and a 3D model explorer. The on-line project team workspace is used to create projects, find and store documents, drawings, models and data, and it enables the project team to share their engineering designs and related files in a secure manner. Anyone with a Web browser and an on-line Internet connector can access the project team workspace through authorisation and authentication. The meeting centre can harness the scheduling of on-line meetings and collaboration in real-time on engineering applications. The 3D model explorer provides a platform to visualise, measure, modify, and mark-up 3D models and 2D drawings.

A collaborative system cannot be simply set-up through equipping a standalone CAD system with IT and communication facilities. Due to the complexity of collaborative design activities and the specific characteristics/requirements of CAD systems, it needs some innovations or even fundamental changes in many aspects of the systems, such as architecture design, communication algorithms, geometric computing algorithms, etc. In Table 6.1, some major differences between standalone and collaborative design systems are highlighted. Meanwhile, some relevant notions are differentiated as follows to highlight the features of collaborative design systems.

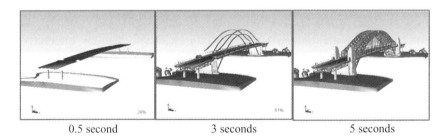

| 0.5 second | 3 seconds | 5 seconds |

Fig. 6.2 Intermediate transmission processes of a design model in RealityWave ConceptStation™ collaboration module.

Table 6.1 Comparisons of current standalone and new collaborative design systems.

Items	Standalone Design Systems	Collaborative Design Systems
Application status	2D or 3D systems are popularly used in product design and development.	Systems are not generally accepted due to weaknesses in interactive capabilities, security of data, real-time and convenient collaboration, etc. Different cultures, educational backgrounds, or design habits of designers make it difficult to organise collaborative design.
R&D status	Mature for geometric modelling. R&D focuses on knowledge-based modelling technologies and application areas, such as mould, die, etc.	R&D focuses on developing new technologies in feature- and assembly-based representations, system architectures, effective distribution/ collaboration algorithms and geometric streaming over the Internet.
Data and system structure	Centrally stored design models. Standalone architecture.	Distributed design models in different geographical sites. Client/server architecture.
Design organisation	Face-to-face discussion, project meeting, review process and whiteboard notification are some common manners for design collaboration. System integration can be achieved through data exchange standards, such as STEP and IGES.	Video conferencing, teleconferencing, electronic whiteboards, WWW forums, live CAD sharing and interoperation are new ways for remote design collaboration. New data exchange, such as XML, HTML and VRML, Internet and Web technology, distributed intelligent, such as multi-agent technologies, are used to establish a distributed integrated system.
Modelling function	2D and 3D modelling capability to support detailed design.	Systems can be classified as visualisation-based collaborative systems, co-design systems, and distributed concurrent engineering design systems. In each level, the requirement for modelling is different.
Market model	Users need to buy and install the packages of CAD systems locally.	New business opportunities and models can be introduced. CAD systems can be designed as remote services for short-term or long-term renting.

(1) Collaborative CAD versus collaborative PDM

Accompanying the evolving process of collaborative CAD, similar products, namely, collaborative PDM, are in keen competitions. CAD vendors are investing large amount of resources in this market, such as PTC Inc. with Windchill™ (www.ptc.com/products/windchill) and Pro/Intralink™ (www.ptc.com/products/windchill/projectlink.htm), UGS Inc. with TeamCenter™ (www.ugs.com/products/teamcenter), IBM Inc. with Enovia™ (www.ibm.com/software/appliations/plm/enovia) and SmarTeam™ (www.smarteam.com), etc. Traditionally, CAD and PDM are regarded as two primary pillars to uphold product design. CAD is a design-centric tool to provide a platform for embodying geometric and engineering design models and drawings, and PDM is a process-centric tool to streamline the information communication and coordination among design departments through storing, managing and exchanging design models and processing information. As collaboration is becoming increasingly more frequent in the today's globalised manufacturing industrial landscape, the current common direction of CAD and PDM is to be collaboration-oriented, and facilitate the reliable, secure and efficient data transfer across the enterprise boundaries. The research issues, challenges and functions towards this direction are becoming overlapping, and a significant trend is to integrate them to support an entire collaborative design chain.

(2) Distribution versus collaboration

A collaborative CAD system needs two kinds of capabilities and facilities: distribution and collaboration. These two terms emphasise the different aspects of a system: physically, the former focuses more on the backbone infrastructures that are used to link dispersed design systems for geographical expansion, and, functionally, the latter associates and coordinates individual systems to fulfil a global design target and objective. In the aspects of enabling technologies, distribution is more fuelled by the development of IT, such as J2EE, .Net, Web, HTML, XML and Web service technologies, and collaboration is more driven by the development of logical and intelligent coordination mechanisms to facilitate human-human/human-computer relationships. Collaboration must not only augment the capabilities of the individual specialists, but also enhance the ability of collaborators to interact with each other and with computational resources [Wang, *et al.*, 2002]. However, although having different focuses, they are closely inter-related and complementary. A collaboration mechanism needs the specific design of the distributed architecture of a system to meet the functional and performance requirements, such as sharing diverse and complex forms of information, supporting a multi-disciplinary design team and integrating heterogeneous application services [Li, *et al.*, 2005(a)].

Synchronous and asynchronous communications are two primary modes to bolster collaborative activities. Synchronous collaborative tools enable efficient communication for users to work together at the same time but at different places, therefore, to meet some real-time collaborative requirements. However, the primary drawback of this manner is that it tends to be costly and requires large bandwidth to realise efficient connection and cooperation, and the simultaneous participation organisation and conflicting schedules are quite challenging. Meanwhile, more complex IT infrastructures and collaborative functions are required accordingly. Asynchronous tools enable communications and collaborations to happen in the "different time-different place" manner. This manner provides the convenience for users to work together based on their own schedules and time with instantly accessible resources and less bandwidth requirements, and some histories of the interactions of a working group can be captured and recorded as backup information. The primary drawback of this manner is that users might experience the impersonal feelings. Therefore, due to the characteristics of synchronous and asynchronous communications, they are selectively used in different collaborative design activities to meet the particular requirements of participants.

(3) Horizontal collaboration versus hierarchical collaboration

According to the functions and roles of the users participating in a collaborative design activity, a collaboration product development system can be organised in either a "horizontal" or a "hierarchical" mode. The horizontal collaboration emphasises on collocating a design team from the same or different disciplines to carry out a task systematically. The hierarchical collaboration can establish an effective communication channel between the upstream design and the downstream manufacturing simulation tools, and it can enrich principles and methodologies of concurrent engineering to link diversified engineering tools dynamically.

Due to the different levels of collaborations and interactions between users, the horizontal collaboration can be further categorised into two types, namely, visualisation-based collaboration and co-design collaboration.

Visualisation-based collaboration has the advantage of facilitating collaborative and distributed conceptual design or product pre-view/re-view. In such an environment, a multi-disciplinary team involving a manager, designer, process planner, customer, etc., can be formed to look at or review the same visualised design model, which is often steered by a chief designer. Net-meeting, email services, and mark-up tools on the design model are some primary ways to facilitate the collaboration process. To alleviate the sluggish transfer of the large-volume design models over the Internet, concise 3D formats for Web applications, such as VRML, have been launched to simplify the models as triangles for visualisation purposes. Under this collaboration, the communication can be maintained through either an asynchronous manner or a synchronous manner.

The more interactive design collaboration (i.e., co-design collaboration) for a conceptual or detailed design requires more complex coordination and organisation among the users. Co-design can be conducted either asynchronously or synchronously. An asynchronously collaborative activity can be organised in a hierarchical assembly structure, through which a chief designer outlines the assembly configuration and the detailed component design tasks are assigned to individual designers to be carried out separately. Managements, coordination and project review of tasks, which can be assisted by some advanced commercial products, such as Microsoft Project™, are vital to the whole process. A synchronous collaborative activity is conducted in a way such that a group of designers are dedicated to the same task actively. Teamwork techniques, such as user commitment, roles and responsibilities, are crucial to guarantee this simultaneous co-design activity. Two scenarios for an asynchronous and a synchronous co-design are shown in Fig. 6.3 and Fig. 6.4 respectively.

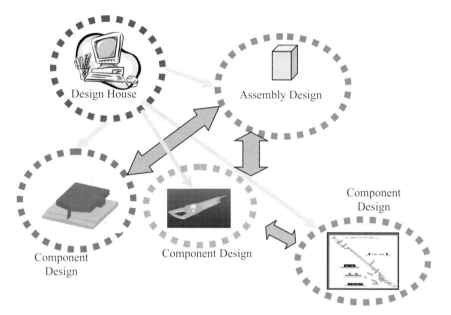

Fig. 6.3 An asynchronous co-design environment.

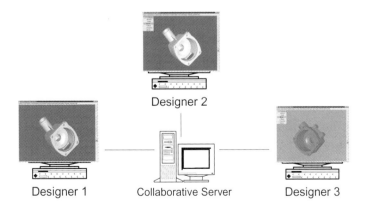

Fig. 6.4 A synchronous co-design environment.

From horizontal collaboration to hierarchical collaboration, with the increased complexity of interactions and system infrastructures, the enabling information technologies are migrating. For instance, it is moving from client-side programming to server-side programming, and further to more complex enterprise integration infrastructures, such as CORBA, Java RMI, agent-based technologies, etc., as shown in Fig. 6.5.

In the following content, the research works and developed systems are surveyed based on the aforementioned collaboration manners, i.e., horizontal and hierarchical design collaborations. The horizontal collaborative systems are summarised from two aspects: (1) visualisation-based systems, which provide a light-weight approach for users to participate in collaborative design through visualising, annotating and inspecting design models in a Web or a CAD environment, and (2) co-design systems, which support interactive co-design functions as a teamwork. In each category, their underlying collaborative mechanisms, distributed architectures and IT implementation strategies are discussed.

6.2 Visualisation-based Collaborative Systems

Visualisation-based collaborative systems have been used to support visualisation, annotation and inspection of design models to provide assistance to the collaborative design activities. The systems are either generically plugged-in a Web browser or added-on viewers in some CAD systems. Some systems and their functions are listed in Table 6.2.

The Web is one of the most popularly Internet tools used to provide a light-weight and an operation system-independent platform for users to search, browse, retrieve and manipulate information disseminated and shared remotely. A visualisation-based collaborative system contained in a Web browser can dynamically share and update 3D models through a standard communication protocol, i.e., HTTP in an Internet environment, to facilitate an on-line team to take on design discussion, product review, design remark and customer survey to enhance collaborative new products and conceptual design. Application services in product design, process planning, engineering analysis and simulation, can be conveniently embedded in the Web as Application Service Providers (ASPs) for remote invoking and manipulation. Considering the requirements of these systems in the Internet and Web with limited bandwidth capability, research has been carried out in the development of light-weight 3D standards and 3D streaming communication. Due to the dominance and effectiveness of the Web technique in the visualisation-based collaborative systems, the following discussions are focused on this perspective.

6.2.1 *3D representations for Web applications*

To deliver and manipulate interactive 3D objects effectively in the Web, some concise formats, such as VRML, X3D (www.x3d.org), W3D (Web 3D) (www.macromedia.com) and MPEG-4, have been launched and the geometry of 3D CAD models can be represented as visualisation-used triangular meshes and trimming lines. VRML is fundamental for these standards to represent geometric elements and scenes, while X3D and MPEG-4 are extended to support VRML and video/audio applications in compressed binary formats, respectively. Formats such as OpenHSF (www.openhsf.org) and XGL/ZGL (www.xglspec.org) are equivalent to the VRML standard in function as they define data for effective 3D streaming transmission through providing functions in data compression, mesh simplification and object prioritising. Most of the above formats are generic and they are not suitable for representing complex CAD models since they lack feature and assembly structures to organise information. The trend in this area is to support and provide complex engineering data and the attributes, advanced streaming and compression formats, strong interoperability and cross-platform capabilities. Some characteristics of VRML, MPEG-4, X3D and OpenHSF are discussed next.

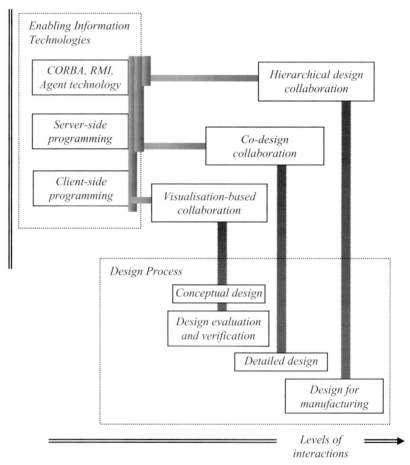

Fig. 6.5 Different design processes with different collaborations and IT
implementations.

Table 6.2 Visualisation-based collaborative systems.

Products	Characteristics and Functions	Data Distribution
Cimmetry Systems Autovue™	(1) A viewer for part and assembly models. (2) View, mark-up, measure, explode, cross-section, etc.	3D streaming
ConceptWorks ™	(1) An add-on viewer to SolidWorks. (2) View and mark-up.	3D streaming
Actify SpinFire™	(1) A viewer for part models. (2) View, cross-section, measure, grid and ruler.	Download
SolidWorks eDrawing™	(1) A viewer for native or simplified SolidWorks files. (2) View, mark-up, measure, 3D pointer, animation.	Download
Centric Software Pivotal Studio™	(1) A base platform to provide a workspace manager, a project organiser and a viewer for part models. (2) View, mark-up, video/audio conferencing, chat.	Download/ 3D streaming
Hoops Streaming Toolkit™	(1) A toolkit to provide 3D streaming APIs. (2) BaseStream class library, advanced compression, attribute (color, texture) support, object prioritization, etc.	3D steaming
RealityWave ConceptStation ™	(1) A VizStream platform, which consists a server and a client. (2) View, mark-up, message.	3D streaming
Autodesk Streamline™	(1) A platform based on the VizStream. (2) View, measure, bill of materials.	3D streaming

(1) VRML

VRML is an international standard scene description language that defines 3D shapes and scenery (or "world") on the Internet. The history of VRML started since the late 1994 and has gone through different versions, namely VRML 1.0, VRML 1.0c, VRML 2.0 and VRML 97.

Basically, VRML 97 is the ISO standard version of VRML 2.0. Essentially, VRML files describe the 3D scenes in terms of objects, operations, and properties of the scenes. It has the advantage of being written in a text format, so that anyone with the desire to change or read the model file can do so with ease. The triangular mesh is the most popular choice of polygons for representing the mesh of a VRML model even though other types of polygons, such as quadrangle and hexagon, can also represent it. The basic geometry information is represented by the contents in *point* (providing the 3D coordinates of vertices) and *coordIndex* (providing the vertex indices in a certain sequence to compose the triangular meshes).

Some characteristics and functions of VRML can be summarised as follows:

- Primitive 3D objects, such as boxes, cubes, spheres and cones, are supported and complex objects can be also supported through extrusion operations.
- Textures can be wrapped to objects.
- Light sources and shading techniques are allowed to alter the appearance of VRML content.
- URL hyperlinks are supported to allow VRML objects to be linked to Web pages or other VRML scenes.
- VRML allows the definition and reuse of objects in programming, such as Java, JavaScript and ECMAScript. Meanwhile, the prototype concept allows custom objects and behaviours to be created and reused as needed.
- Sensors are embedded to allow VRML objects and scenes to sense and respond to the passing of time and user activities.
- Background colours, images and sounds are supported.

A VRML plug-in is needed in a Web browser when a VRML file is displayed. The plug-in parses a VRML file and render the 3D contents into the Web browser while handling user interaction and navigation. Several popular plug-ins are listed in Table 6.3.

Table 6.3 Popular VRML plug-ins

Plug-ins	Companies	URLs
Contact™	Blaxxun	www.blaxxun.com
Cortona™	Parallel Graphics	www.parallelgraphics.com
Cosmo Player™	Computer Associates	www.cosmosoftware.com

The major disadvantages of VRML include large files sizes, no built-in compression and no streaming technology. Hence, the download and display speeds are not satisfactory.

(2) X3D

X3D is a major upgrade from VRML and retains a backwards compatibility with a huge base of available 3D content. It is being developed under the Web3D Consortium's (www.web3d.org) standardisation process that provides full and open access to the specifications and eventual submission to ISO for ratification, to provide long-term stability for Web3D content and applications. The main motivation for moving to X3D is a more light-weight representation, hence eliminating the need to download a heavy browser. In X3D, a browser is downloadable with content. Several popular X3D plug-ins for Web browsers are listed in Table 6.4.

Table 6.4 Popular X3D plug-ins

Plug-ins	Features	URLs
Xj3D	An open-source X3D web browser based on Java and XML	www.xj3d.org
Flux™	An ActiveX plug-in	www.mediamachines.com
BS Contact™	Support 3D stereo	www.bitmanagement.de

Major features of X3D include:

- X3D is XMLised. XML is the most popular new generation Internet application language. Nodes in X3D are represented in XML tags so as to take full advantage and potentials of XML on the Internet.
- X3D utilises an open profile/components-based architecture enabling custom-crafted scalable implementations. It has a layered and componentised architecture that enables extremely compact 3D clients. These clients can be extended with plug-in components to create standardised profiles with the functionality to meet the demands of sophisticated applications.
- X3D incorporates numerous advanced 3D techniques including advanced rendering and multi-texturing, NURBS surfaces, GeoSpatial referencing, Humanoid Animation (H-Anim), and IEEE Distributed Interactive Simulation (DIS) networking.

(3) MPEG-4

The bulky sizes of VRML files hinder their effective transmission over the Internet. MPEG-4 has been proposed to define a binary compression format, i.e., BIFS (Binary Format For Scenes), to encode VRML in a binary representation. Therefore, a BIFS file is often 10 to 20 times smaller in size than its VRML equivalent. As the major objective of adopting MPEG-4 is for multi-media applications, MPEG-4 supports media mixing and audio composition, so that it can easily mix with rich forms of media, video and audio.

(4) OpenHSF

VRML, X3D, MPEG-4 and other formats, such as ZGL and W3D, are designed for more general and light-weighted uses of visual information for the Web and Internet applications. However, they are not suitable for engineering data types, such as complex assemblies and associated 2D drawings. OpenHSF has been proposed by Hoops3D Inc. to handle specific visualisation requirements of mechanical CAD and architecture/construction software. Several popular plug-ins for Web browsers or standalone viewers are listed in Table 6.5.

Table 6.5 Popular OpenHSF plug-ins or viewers

Plug-ins/viewers	Features	URLs
HOOPS stream control™/ stream plug-in™	IE/Netscape plug-ins	www.hoops3d.com
ParaHOOPS 3D part viewer™	A viewer for Parasolid and OpenHSF files	www.hoops3d.com
ACISHOOPS 3D part viewer ™	A viewer for ACIS and OpenHSF files	www.hoops3d.com

Some advantages of OpenHSF include:

- It supports engineering geometry and engineering attributes. The numerous geometric primitives supported by OpenHSF include basic elements, such as vertices, arcs and circles, NURBS, multi-byte text, images, cutting planes, etc. The attributes tailored by OpenHSF include textures, transparency, colour, line styles, marker symbols, face and line patterns, per-vertex colour and iso-lines for CAE, etc.
- It supports 3D streaming and compression. OpenHSF is a stream-capable format with support for streaming concepts, such as multi-resolution objects (Level-of-Details) and file ordering. A compression algorithm is embedded for efficient transmission over networks.
- Open format and interoperability. OpenHSF is an open, published format, and users can read and write OpenHSF files without consideration of IPs. OpenHSF has interoperability to support data exchange with some leading CAD and CAE vendors.

6.2.2 *System architectures and implementation strategies*

A visualisation-based collaborative system proposes a two- or three-tier client/server architecture. Fig. 6.6 shows a three-tier architecture, which includes client(s) (a Web browser and/or an application client), server(s) (a Web container and/or an application server) and database server(s). If no database server is used, the system is simplified as a two-tier architecture (e.g., the grey part in Fig. 6.6).

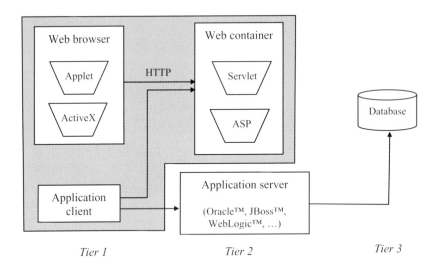

Fig. 6.6 A multi-layered client-server architecture.

Java Applet and Microsoft ActiveX technologies are widely used for developing Web-based clients, and some services written in Java Servlet, Microsoft .Net ASP or CGI (Common Gateway Interface) are deployed in the server side to provide support and system maintenance [Lee, *et al.*, 1999; Shyamsundar and Gadh, 2002; Chen, *et al.*, 2003; Zhang, *et al.*, 2004]. Java3D is a widely used programming APIs (Application Programming Interfaces) launched by Sun Inc. to enable visualisation-based manipulations of 3D objects and scenes, and a client Applet can be developed based on Java3D to build, render and control 3D objects for Web-based collaboration.

(1) Java3D

Java3D is sometimes misunderstood as a Web-based representation scheme equivalent to VRML in function. However, they have fundamental differences although there are close relationships between them. According to the tutorials of Java3D and VRML, "Java3D is a high-level programming APIs for 3D graphics rendering. The code must be compiled to move it to an executable form. VRML is a text-based modelling language that is interpreted dynamically from the source files. A VRML browser can be implemented using Java3D, but one cannot create Java3D applications with VRML". In other words, VRML is "static" and consists of a series of text presented in certain formats, and Java3D is "dynamic" as it is a Java programming language developed specially for visualising and manipulating 3D models. Usually, VRML is used as the input information for a Java3D program.

Java3D defines several classes that are used to construct and manipulate a hierarchical scene graph to control viewing and rendering. A typical scene graph is shown in Fig. 6.7. In the graph, the VirtualUniverse object provides a grounding for scene graphs, and the Locale object defines the origin of its attached branch graph in coordinates. Geometries, appearances of the geometries (e.g., colour, texture, material, etc.), the relevant transformation information (e.g., positions, orientations and scales of the geometries), and behaviours (e.g., keyboard and mouse manipulations for the geometries in the scene) are specified in a BranchGroup node. Some features of Java3D are summarised as follows:

- A rich set of 2D and 3D objects and behaviours. A number of geometric 2D and 3D objects, and their attributes are defined. Behaviours provide the means for animating objects, processing keyboard and mouse inputs, reacting to movement, and enabling and processing events.

- 3D geometry compression and support of Level-of-Detail (LOD) objects. A binary geometry compression format (a generalised triangle strip format) is utilised in Java3D, both as a run-time in-memory format for describing geometry, as well as a storage and network format. The LOD method can store and display different layers of detailed objects.

- Wide variety of file formats. Run-time loaders are supported to accommodate a variety of file formats, such as vendor-specific CAD formats, interchange formats and VRML.

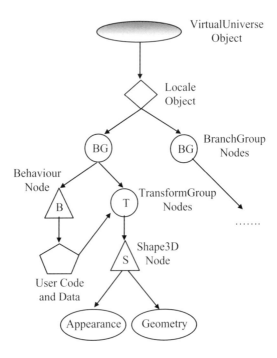

Fig. 6.7 The scene graph of Java3D.

In a Web application, a Java3D class is usually instantiated in a Java Applet program, and an HTML file is created to hold the Applet and rendered in a Web browser (Fig. 6.8).

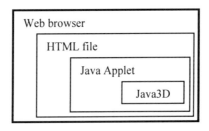

Fig. 6.8 A Java3D class is contained in a Web.

(2) System implementation strategies

There are two implementation strategies for a Web-based visualisation system, i.e., the client-side programming paradigm, and the hybrid client- and server-side programming paradigm. Different functional strategies have been developed to support collaborative activities based on these strategies.

Java Applet or Microsoft ActiveX, which are client-side programming techniques, can be embedded in an HTML file and downloaded from a server to be executed in a client (Fig. 6.9). This paradigm provides an efficient handling of the frequent interactive operations on models by the users locally; therefore, many works have adopted this strategy to develop visualisation systems.

Fig. 6.9 The client-side programming scenario.

Roy and Kodkani [1999] proposed a Web-based framework to support collaborative product design with various levels of abstracted product models. In their framework, a designer creates the model using a conventional CAD package and uploads this model to a database system in the server. The product model is decomposed at face and feature levels to address different applications and represented in VRML formats and HTML pages. In their framework, VRWeb (www2.iicm.edu/vrweb), which is an open-source program based on Java Applet and Java3D, is employed to display the relevant design models. Kan, *et al.* [2001] designed a Web-based graphical synchronisation engine to support three collaborative functions for multiple users. The first function specifies the status of users of a collaborative working session. When a new user logs in or out of the system, the "avatar" (3D object that represents the user) that is added or removed from the collaborative environment can be synchronised in all the users' screens. The second function synchronises the design model that has been added by a user to the other users' screens. The third function allows users to share their viewpoints, which means a user can request and switch to another user's viewpoint so that they can have the same viewing perspective. The proposed graphical synchronisation engine is built based on the above paradigm with a Java Applet and a VRML browser (Cosmo Player™ (www.cosmosoftware. com)). In the Web-based product information sharing and visualisation prototype developed by Zhang, *et al.* [2004], product structure trees and product data master models are used to capture and manage essential and inter-related product information/properties, such as design data, material properties, geometric and topological structures, finite element analysis and optimisation, process planning, scheduling, manufacturing, purchasing and supply management. 3D product models are shared in the Web through Java Applet and Java3D. ActiveX was used to facilitate the establishments of Web-based DFX (Design For X) tools and a morphological chart-enabled collaborative conceptual design environment [Huang, *et al.*, 1999; Huang, 2002].

However, this paradigm has the following limitations for setting up fully functional collaborative systems:

- This paradigm is composed of a series of repeatable request-downloading processes of static HTML pages. A server plays the storage and management functions of the executive codes and HTML pages, and the life-cycle of each collaboration process between a client and a server is a uni-directional information flow process. This characteristic hinders the efficient broadcasting of design changes in a client to other clients, which is a common situation in interactive collaboration.
- In a collaborative development environment, sometimes, it is necessary to monitor and coordinate the utilisations of functional modules deployed in the server. However, under this paradigm, the server loses the control of the relevant HTML pages and executive codes when a downloading process finishes. This is undesirable in some situations such as time-dependent charging services.

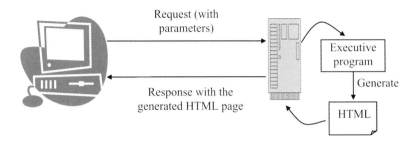

Fig. 6.10 The server-side programming scenario.

Server-side programming, such as J2EE (including Servlet, JSP-JavaServer Page), Javaspace, CGI, Microsoft .Net ASP, is an emerging and promising technique to overcome the above drawbacks through executing programs on a server to enhance its capabilities. Fig. 6.10 shows this scenario. However, this approach is quite sluggish if frequent interactive operations on 3D models, such as rotating, zooming or changing viewpoints, are needed. Considering the CAD characteristics and the specific requirements for interactive and real-time operations, a hybrid paradigm of combined server- and client-side programming is becoming more popular to take advantage of their complementary functions. As such, Applet or ActiveX is used to provide an efficient handling of interactive visualisation-based manipulations on 3D models by the users locally. Servlet, .Net ASP, CGI, RMI, etc., are used to monitor and manage collaborative activities and establish the effective dual-communication between the clients and the server. This hybrid paradigm has been implemented by the more recent works to incorporate advanced collaborative functions.

Gera, *et al.* [2002] developed multi-user groups for conceptual understanding and prototyping system (MUG) to create a shared design workspace to facilitate design process. 3D design models, design structures and function descriptions are some sharable design semantic information for a group of designers to evaluate. Java3D, VRML and X3D are used on clients, and Javaspace services are instanced for mapping clients' performance operations and entries and managing the invocations of methods on the server to respond to clients. CyberCAD [Tay and Roy, 2003] provides a 3D CAD sharing and collaborative design environment. Java3D and the relevant client-side programming techniques are adopted. To maintain the complex inter-relationships and multi-media communications between designers and design models, a server-side programming technique, i.e., Java RMI is deployed to realise the interoperations of a client and server, so as to update the changed design efficiently in the working designer group.

With more advanced and enhanced collaborative functions, systems are evolving to become more dedicated to support detailed design and users are equipped with facilities to actively involve them in design tasks. This situation will be discussed next.

6.3 Co-design Collaborative Systems

In a distributed detailed design process, a co-design activity requires more active participations from a design team. There are two co-design activity organisation paradigms, namely, synchronous co-modelling/co-modification design, and asynchronous assembly-based co-design. To satisfy the requirements, different system infrastructures are specified.

In a synchronous co-modelling/co-modification paradigm, each user is enabled to participate in design collaboration synchronously with modelling and modification capabilities. During iterative design sessions, changes imposed by a designer can be communicated to other project participants through concurrent sharing and merging these changes with the design models of other designers. Therefore, suitable coordination and synchronisation mechanisms are crucial to schedule a design activity in parallel and ensure no conflict arises during this real-time and iterative design process. Meanwhile, as real-time data sharing, which is an essential requirement to ensure the collaboration, is almost impossible due to the huge design models and limited bandwidth of the Internet, a new feature representation scheme is actively being explored.

In an asynchronous assembly-based paradigm, a co-design activity is centrally coordinated at an assembly level. Assembly constraints are encapsulated as interfaces to provide different designers with platforms to cooperate, which can ensure sub-assemblies and components allocated to individual designers are compatible with each other. Although real-time sharing is not a must, an optimised representation strategy for assemblies to simplify data and avoid the sluggish transmission is still desired. Meanwhile, a propagation mechanism for changes in a sub-assembly or component to the entire assembly structure is imperative to maintain the assembly consistency. Hence, the following three aspects have to be investigated:

- An effective system architecture based on the available IT infrastructures, such as client/server, peer-to-peer and Web service.
- New feature and assembly representations and schemes to optimise data sharing, transmission and management in the distributed environment.
- Effective team organisation, coordination and negotiation to ensure the success of a collaboration process and drive down the design cycle time.

6.3.1 *System architectures*

The underlying architectures of the systems to enable co-design can be classified into the following three types (Table 6.6):

- Communication server + modelling client (Thin server + thick client)
- Modelling server + visualised-based manipulation client (Thick server + thin client)
- Application or service sharing (Peer-to-peer)

(1) Communication server + modelling client

In this scenario, clients (a Web-based client or application client) are equipped with CAD functions and some communication facilitators. A server acts as an information exchanger to broadcast CAD files or commands generated by a client to other clients during a collaborative design process. Some developed systems that use this architecture include CollabCAD™ (www.collabcad.com), IX SPeeD™ (www.impactxoft.com/products/suiteV5.asp), Nam and Wright [1998], Qiang, *et al.* [2001], Bianconi and Conti [2003], etc.

Table 6.6 Three mechanisms for distributed systems.

Mechanisms	Illustrative diagrams
Communication server + modelling client	
Modelling server + visualisation-based manipulation client	
Application or service sharing	

From the architecture perspective, this scenario is similar to that of a Web-based visualisation system in the client-side programming paradigm. However, more complex modelling and modification functions are supported here, and team organisation and synchronisation are necessary to ensure design data consistency and leverage co-design activities. Some of these developed systems are based on commercial CAD systems to implement the modelling functions (CATIA™ (www.catia.com) open architecture in IX SpeeD™; Unigraphics™ (www.ugs.com) in Qiang, *et al.*' system; SolidWorks™ (www.solidworks.com) and SolidEdge™ (www.solidedge.com) in

Bianconi and Conti's system; etc.) while others are based on solid modelling kernels (Open CASCADE in CollaCAD™, Allias in Nam and Wright's system, etc.).

CollabCAD™, which was developed by National Informatics Centre of India, allows multiple designers to access and work on the same part concurrently, and audio and video facilities are equipped to provide multi-media collaborative functions. Designed parts and commands can be uploaded in a server and broadcast to all the other designers who are participating in this design session to update the design. IX SPeeD™ supports an innovative "sharing and merging" technique to help designers synchronise the changed design with their working models. Different "media" are used to exchange information among co-design systems. Nam and Wright [1998] updated a design model that has been developed by a designer to all the other design systems (clients), while the method proposed by Qiang, *et al.* [2001] records all the commands made in a workstation and transfers these commands to another workstation in the network to rebuild the design scenario to achieve simultaneous co-design. To support a heterogeneous co-design environment, in which different CAD clients are employed, Bianconi and Conti designed a neutral exchange format based on XML to encode design semantics of CAD systems. As such, the native representations of different CAD platforms are filtered, and inter-operations between different CAD systems can be realised for co-design.

(2) Modelling server + visualisation-based manipulation client

The hybrid paradigm of the Web-based visualisation systems enhances a server so that it can establish more effective collaborative functions and fully realise controllable remote services. In this scenario, the data structures in clients are light-weight and they primarily support visualisation and manipulation functions (such as selection, transformation, changing visualisation properties of displayed parts, etc.). The main modelling activities are carried out in a common workspace in the server side. A thin/thick representation in the client/server respectively has been proposed to enhance the performance of the system effectively. The developed systems include Alibre

Design™ (www.alibre.com), OneSpace™ (www.onespace.net), Lee, *et al.* [1999], van den Berg, *et al.* [2002], Li, *et al.* [2004(a)], Wu and Sarma [2001, 2004], etc.

The Alibre Design™ system features real-time team modelling capabilities and the ability to share data via distributed repositories. The real-time team modelling offers an environment for geographically dispersed engineers to jointly view, annotate, and edit a design. The system architecture is three-tiered: a light-weight Web-based client interface, application design server to carry out major geometric computing, and a database repository server to manage design and processing information. Based on the Web service architecture, OneSpace.net™ provides 3D co-design explorer, and some collaborative facilities similar to those existing in other co-design systems, such as organised projects, secure messaging, presence awareness and real-time on-line meetings. To resolve the problem of sluggish manipulations during a co-design process under this scenario, Lee, *et al.* [1999], Wu and Sarma [2001, 2004] and Li, *et al.* [2004(a)] proposed several algorithms to differentiate changed and unchanged data, and only transfer the changed data in the design team for synchronisation so as to reduce the traffic volume.

(3) Application or service sharing

Different from a client/server architecture, in which a network is hierarchically managed by a centralised server computer and information is made available to the client computers, a peer-to-peer architecture allows a group of computers to be connected with equivalent responsibilities to pool their resources and decentralise the management. Based on the peer-to-peer architecture, individual computers and executive applications are transformed into shared resources that are accessible from each of the computers, as shown in Fig. 6.11. Based on the peer-to-peer architecture, Begole, *et al.* [1997] and Inventor™ (www.autodesk.com) support the sharing and manipulation of services or modules of a system by other systems. For instance, for the Inventor™, an Microsoft Netmeeting tool, which provides video conferencing, whiteboard, chat, file transfer and windows based application sharing tools, is embedded for co-design communications and application sharing. During a collaborative design process, an Inventor system can work jointly with another system through multi-media communication, or take over the control of the second system for remote manipulation.

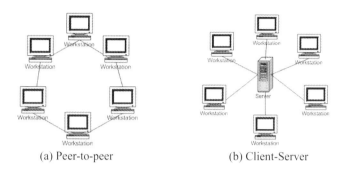

(a) Peer-to-peer   (b) Client-Server

Fig. 6.11 Scenario for two network architectures.

(4) Comparisons of the three architectures

Considering the characteristics of collaborative CAD systems, the above three architectures show potentials in different aspects. The implementation of the first architecture is quite straight-forward as compared to the other two architectures. Through equipped with a communication facility, standalone CAD systems can be conveniently re-developed as design clients and linked together by a server with functionalities for information exchange and collaboration coordination. This architecture can effectively meet the requirements of CAD design for real-time interactive operations since most of the geometric computing for modelling and modification is carried out locally in the clients. At the same time, it can support heterogeneous modelling systems in clients and a neutral information exchange format, for instance, XML, can be designed for communication in the environment [Bianconi and Conti, 2003]. However, this architecture lacks adaptability. If a new user is added in the environment, the whole CAD system has to be added and configured. Meanwhile, this architecture is difficult to be migrated to a Web application.

The second architecture is becoming more popular as it brings forth a new business model – ASP. With such an architecture and ASP-enabled collaborative CAD, small and medium enterprises can rent on-line high-end collaborative CAD systems, so that they are able to participate and co-operate in the design process with large design companies. Through this manner, renting on-line high-end CAD systems that are running on CATIA™ and Unigraphics™ systems have now become affordable for small and medium enterprises, and not just only for big companies. The scalability of a system can be enhanced since it is convenient to add new seats in the distributed system. The problems that hinder it are the increased implementation difficulty and the sluggish system speed due to the vast amount of information exchange across networks.

The third architecture employed the peer-to-peer computing manner. Services of a peer with CAD functions can be manipulated by another peer. The architecture has high performance for point-to-point communication and collaboration, and provides a more flexible manner to integrate individual systems in the environment. However, it is not suitable for a large group of users to work together due to the restrictions of the peer-to-peer computing. Another obstacle is that the coordination among the decentralised systems is more complicated comparing to the other two architectures.

6.3.2 *Design coordination and team management*

In Table 6.7, four collaborative CAD systems are summarised. In these systems, the team management and design coordination functions are the crucial components for establishing a well-organised team to conduct a collaborative design task and avoid design conflicts. Successful coordination and team management require effectiveness in the following several aspects:

- Awareness of collaborative partners
- Sharing and communication of data, knowledge and information
- Negotiation and conflict management for cognitive synchronisation/reconciliation
- Working session and coordination mechanism

Table 6.7 Co-design systems and their team management/design coordination.

Systems	Collaborations	Functions & information distribution	Supported systems & formats	Modelling functions	OS platforms
Alibre Design	*Team design sessions* A design session can be used to organise a virtual team to design 2D and 3D models simultaneously. *Repository* Through repository, users' models can be securely shared and accessed. *Message Centre* It can support message sharing among users.	• Modelling • Mark-up • Annotation • View • Text and voice chat • Directly transferring CAD models • Client/Server communication	• Pro/E • UG • CATIA • AutoCAD • STEP • IGES • SAT • STL	• 2D sketching • Dimensioning • 3D modelling • Assembly modelling • Bill Of Material (BOM)	• NT • 95/98 • 2000 • ME • XP
CoCreate OneSpace	*3D personal collaboration* Through this service, up to two other users can be invited to an on-line meeting. Meeting users can view and mark-up 2D or 3D models. *Model manager* It can store and share users' models through a database, and specify who has permission to read and modify design work. *Project data manager* It can organise 2D or 3D project files in a database and helps track of document version and history.	• Modelling • Mark-up • View • Netmeeting • Integration with PDM • Directly transferring CAD models • Client/Server communication	• I-DEAS • CATIA • Pro/E • UG • SolidWorks • STEP • IGES • IDF • EMatrix • Word • PDF • zip	• 2D sketching • Dimensioning & tolerancing • 3D part modelling • Assembly modelling • CAE analysis • Inspection • Mould base • Mould design • Mould flow analysis • Sheet metal • Geometric simplification • Surfacing	• Unix • NT • 95/98 • 2000 • ME • XP

Table 6.7 Co-design systems and their team management/design coordination (cont'd).

Systems	Collaborations	Functions & information distribution	Supported systems & formats	Modelling functions	OS platforms
Autodesk Inventor Collaborative Tool	*Application sharing* Microsoft Netmeeting tool is embedded into Inventor systems to organise co-design activities. An Inventor that has the "control baton" can control and manipulate another remote Inventor system to design and the controlled system is an observer. The "control baton" can be acquired and exchanged.	• Modelling • Netmeeting, Whiteboard • Chat • Directly transferring CAD models • Netmeeting T.120 communication	• Inventor • Inventor compatible files	• 2D sketching • Dimensioning & tolerancing • 3D part modelling • Assembly modelling • CAE analysis • Inspection • Sheet metal	• NT • 2000 • ME • XP
CollabCAD	*Design team* Members in the team can simultaneously design and share 2D and 3D models. *Repository* It can store and share users' models through a database.	• Modelling • Text or voice chat • Directly transferring CAD models • Client/Server communication	• STEP • IGES • VRML	• 2D sketching • 3D part modelling • Assembly modelling • CNC	• NT • 2000 • XP

When collaborations are not possible at the same physical place, the cues of the intra-team relationships, such as face-to-face communication, body language, physical gestures and direct physical contact, are not direct. Awareness of collaborators and their characteristic and activities is necessary for establishing a mutual understanding during collaborative design. Emails, Netmeeting, discussion forums and whiteboards are some usual tools to simulate face-to-face meetings and negotiation to resolve the remote design conflicts and realise cognitive synchronisation [Alibre Design™, OneSpace™, CollabCAD™, Gera, *et al.*, 2002; Tay and Roy, 2003]. Using these tools, designers' on-site information, such as facial expressions and gestures, can be shared to improve the synchronous collaborative effectiveness. Meanwhile, an important mechanism to establish and develop team spirit among remote members is to actively provide constant awareness and feedback of the status and activities of the members who are contributing to the collaborative tasks. Therefore, the information of the collaborators, their current locations, the tasks they are performing and their workloads is important for the coordination of the whole process [Gutwin and Greenberg, 1999; Xie and Slavendy, 2003]. For instance, if a collaborator is aware of the workload of his partners, he might work and share some tasks with partners who have a low workload and would take some responsibilities voluntarily or passively from the over-loaded partners. Furthermore, some efforts [Huang, *et al.*, 1999; Lang, *et al.*, 2002] have been made to inspire individual dedications through providing mechanisms to record and rate member contributions and share these ratings across projects. These mechanisms foster improved cooperation as individual assistance would be reciprocated and their reputations can be set-up gradually. More advanced studies on psychological and social aspects of collaborative awareness are actively being explored to achieve a balance between the privacy and transparency of individual design information and activity.

Sharing and exchanging design data, knowledge and information dynamically is a vital way to realise collaboration. Through messages in text, video or audio, designers can communicate with each other to exchange design ideas. More advanced functions are being developed to generate the minutes or video transcripts of meetings, caching and indexing of emails. However, it is not easy to carry out efficient or even real-time sharing of large-volume engineering design data and models. Some optimisation strategies for this problem are discussed in Section 6.3.3. In Table 6.8, some functionalities of a centrally or remotely located working group are listed [Tay and Roy, 2003].

Table 6.8 Functionalities for collaborative activities centrally and remotely located.

	Same place	Different place	
Same time	Meetings Whiteboards Face-to-face communications	Video conferencing Teleconferencing Screen saving Electronic whiteboards WWW forums Live CAD sharing and interoperation	Synchronous collaboration
Different time	Mailboxes Shared folders Bulletin boards Documents Meeting minutes	E-mails Workflow Form flow Messaging Routing and notification	Asynchronous collaboration

Negotiation is an important way of formalising and implementing mediation and facilitation functions among designers to handle design conflicts. Earlier coordination mechanisms [Mintzberg, 1983] used direct supervision to allow an individual designer or agent to issue instructions and monitor the execution of the tasks. Working process, output and skills were required to be standardised to reduce the mutual communication and information exchange. Negotiation improves the rigidity of these mechanisms by introducing more flexible and rational conflict arbitration mechanisms. The negotiation model proposed by Case and Lu [1995] includes an identification of conflicts, dissemination of information about conflicts, exchange of rationale and recording of conflict resolution. The process is initiated through an identification of a conflict event. A set of users who have shown an interest in the artefacts involved in the conflict event is notified and a deadline is set for negotiations. After receiving a notice of an event, each user can express an opinion about the desired outcome of the conflict, which will then be sent to the arbiter. Each user is kept informed about the status of the negotiations and has the option to change his opinion at any time until the deadline is reached or the conflict is resolved. If the conflict cannot be resolved through the negotiations and the time reaches the resolution deadline, the arbiter will select one of the opinions, label it as the resolution, and broadcast it to all the interested users. Wong [1997] proposed four conflict resolution methods, which include enquiry, arbitration, persuasion and accommodation. In the enquiry method, designers or agents find and collect the underlying data and beliefs about a conflict. They then resolve the conflict by appealing to the retrieved data and beliefs and some shared principles for interpretation. Some arbitration methods are developed based on a fair social-choice theory to select an outcome from among many competing alternatives. These methods usually include an agenda to contain a series of criteria for judgement. Individual preferences through qualitative means (such as preference relationships) or quantitative means (such as utilisations or probabilities) are first formed from the competing alternative plans according to the criteria. Some procedures are then applied to select an outcome out of the individual preferences. The persuasion method aims at achieving an agreement among the designers. The goal of the enquiry is to settle the conflict through smoothing out the differences, while the persuasion method starts off from a premise that an agreement is not possible and attempts to reshape the agenda to figure out new acceptable solutions. The accommodation method takes into account the history of the decision-making process for conflicts, and some rejected criteria might be reconsidered and included into an agenda as a compromise or compensation. This method is used to address a situation when a former decision differs in what the designers currently desire to rectify the previous action.

A working session mechanism is effective for team management. Each session can be used to organise a collaborative task, and users in the same session can share the design information dynamically [Alibre Design™, OneSpace™, Li, *et al.*, 2004(b)]. Different design tasks can be carried out simultaneously in different sessions. In a session, designers can have different roles, responsibilities and commitment, which situation is similar to a practical working group. For instance, a user can serve as a project leader, team member or supporter. A project leader is responsible for managing a session and supervising the design process, and he/she is authorised to schedule the process and avoid deadlocks during design due to network problems. A design member can carry out the design, and a supporter can provide comments for the design or resources required arising during the collaborative design. For synchronous and asynchronous collaborations, the coordination mechanisms are different. For a synchronous activity, a "control token" mechanism has been utilised to schedule a collaborative design activity [Alibre Design™, Li, *et al.*, 2004(b)]. Each session has a control token, i.e., at any one time, only the user who holds the control token is the active designer and can edit a part; the other users in the same session can only receive the updated information and are observers. The user who is carrying out the editing function can become an observer by transferring his control token to another user. The scenario for this mechanism is illustrated in Fig. 6.12. For an asynchronous activity, commercial software tools, such as Microsoft Project™, have been effectively embedded in some collaborative systems to plan and manage design activities and projects [e.g., OneSpace™]. Windchill ProjectLink™ is launched as an extensive management tool in the Internet for projects, schedules, information and processes. Its major functionalities include project management with milestones, activities, tasks, deliverables and resources, automation and standardisation of key business processes such as new product introduction, six sigma and product quality planning, integration with mechanical and electrical CAD applications, Microsoft Project™/Office™ and the Windows desktop, and operation integrally with the Windchill PDM system. These tools and some prototypes [Saad and Maher, 1996; Fussell, *et al.*, 1998] are complementary and integrated parts of collaborative systems. On the other hand, the handling of dynamic design changes is an important topic in collaborative project management, and some mechanisms have been developed to address this issue. Mori and Cutkosky [1998] proposed an agent-based system to coordinate design changes based on the theory of Pareto optimality. The agents are reactive and can track and respond to changes in the state of the design when any designer changes his model and incurs the conflicts. Zhou, *et al.* [2003] reported an on-line visual workflow to monitor a collaborative design process and design changes. Based on the information monitored, dynamic adjustments of the project schedule can be made on a more accurate and practical basis.

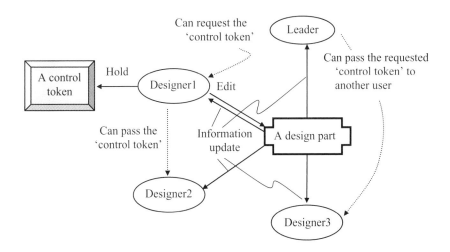

Fig. 6.12 The control process of designing a part through a "control token".

6.3.3 *Optimised feature and assembly-based representations*

A significant problem in these systems is that communication efficiencies are still far from satisfactory when large-sized feature- and assembly-based models are designed collaboratively. To address this problem, some works have been reported recently to optimise or simplify geometric entities of distributed design feature- or assembly feature-based models to accelerate the communication.

Wu and Sarma [2001, 2004] developed an algorithm to incrementally update the B-Rep of a design model based on a cellular representation in a distributed environment. Based on the cells from the segmented B-Rep of a design model, the algorithm can identify and extract regions that have been modified by a designer, and dynamically transmit and embed the modified regions into a B-Rep at another site. Lee, *et al.* [1999] proposed a network-centric virtual prototyping system based on a distributed computing architecture, in which a shape abstracting mechanism was developed to provide a light-weight Abstracted Attributed B-rep (AAB) in clients to represent a feature-based model stored and maintained in a server for concise and transparent communication between the server and the clients over the network. A naming consistency paradigm was established to maintain the interoperability and identification between the geometric entities of the server and the clients during a concurrent design process. The scheme generically names geometric entities that are invariant during geometric processing, such as topological changes and Boolean operations. It associates each model with a *FaceIdGraph* that is updated each time the topology of the model changes, so as to retain the information of how the faces of a model are created, split, merged, trimmed and deleted. Li, *et al.* [2004(a)] developed a distributed feature mechanism to filter the varied information of a working part during a co-design activity to avoid unnecessary re-transferring of the complete large-sized CAD files each time any interactive operation is imposed on the model by a client, so as to enhance the effectiveness of the information communication for co-design activities.

To support collaborative assembly design activities effectively, Shyamsundar and Gadh [2002] developed a geometric representation named as AREP, and a collaborative prototyping system based on this representation to perform real-time geometric modification for components/sub-assemblies in an assembly model. In AREP, an envelop mechanism was designed to simplify some internal geometric structures and entities, which are irrelevant to the assembly constraints, of the designed components separately and collaborated based on the assembly constraints. Points are kept in envelopes to refer to their corresponding detailed entities for further query and retrieval. Chen, *et al.* [2004] proposed an assembly representation for collaborative design, which includes two functional modules, namely a Master Assembly Model (MAM) and a Slave Assembly Model (SAM). The MAM is a complete representation that is stored in the server, and SAM is a simplified version of MAM that is used for visualisation-based manipulations in the client. However, it does not address real-time design modifications in a collaborative design environment. The research direction is towards supporting optimised traffic and real-time feature and assembly design.

6.4 Hierarchical Collaborative Systems

In hierarchical collaborative systems, an upstream design activity is vertically linked with the downstream manufacturing practices. Different from the traditional "sequential engineering", which is a "throwing-over-wall" approach, this collaboration focuses on bi-directional communications and interactions among the designers. Mechanisms for information exchange, system integration, wrapping and invocation of manufacturing services, storing, accessing and maintaining CAD and analysis information securely and conveniently are some crucial research topics to establish an effective collaborative environment.

6.4.1 *Multi-representation and conversion schemes for features*

To support feature-based applications in a collaborative design environment, Gadh and Sonthi [1998] developed a four-level feature representation scheme to address different applications effectively. The representation scheme consists of boundary representation, aggregate geometric abstraction representation, domain-independent geometric abstraction representation and domain dependent features. The objective of this representation scheme is to provide several layers of geometric abstractions and aggregations in a server to respond to different manufacturing applications efficiently. Han and Requicha [1998(b)] and De Martino, *et al.* [1998] separately developed a distributed system consisting of a design-by-feature client and a downstream manufacturing feature recognition client connected through a geometric server. The functions of the geometric server are twofold: first, it is a repository to store features generated by these two clients; and second, it transfers design features in the design-by-feature client to the feature recognition client. The distinction between these two works lies in their feature recognition algorithms. The former uses a hint-based reasoning method depending upon the augmented design features as hints, whilst the latter develops a graph-based reasoning method to work on the geometric models that have been converted from the design feature models. In the above works, changes made in the design-by-feature client can be propagated to the feature recognition client automatically to achieve data completeness and consistency. However, this information flow is uni-directional. If a modification of a design part is required by the manufacturing feature recognition client, it has to be made in the design-by-feature client. This forces a user to think in a way that is not natural to him/her and blurs the functional differences among design and manufacturing. Hoffmann and Joan-Arinyo [2000] proposed a master model scenario to store the shared design information and a multi-way communication mechanism among the design and manufacturing clients. However, this work is limited to some simple types of features and the work is still far from practical applications. This problem can be effectively solved through developing a generic and robust integration strategy to integrate design-by-feature and feature recognition algorithms, to support multiple views of a design model. This has been actively investigated by de Kraker, *et al.* [1997], Jha and Gurumoorthy [2000] and Li, *et al.* [2002].

6.4.2 *Integration mechanisms for distributed systems*

In a collaborative product development environment, design models, analysis and optimisation information, and some intermediate data are actively exchanged and shared. An effective mechanism is imperative to leverage the seamless integration of functional modules that are geographically dispersed, and ensures that correct information is delivered at the right time in the right sequence and order to the right person. Integration mechanisms that have been reported can be categorised into two types: (1) data-centric integration and (2) service-centric integration. Some physical infrastructures based on CORBA, J2EE, .Net and agent-based systems have been proposed to fulfil the relevant functions.

(1) Data-centric integration

Due to the proprietary formats of functional modules, information sharing and exchange is usually hindered in a collaborative environment. Some neutral information exchange standards, such as IGES or STEP, were popularly used in some research works to achieve system integration. However, IGES and STEP data are quite "heavy" in the Internet, and VRML was chosen by some works to represent models in a light-weight mode to be dispatched for remote visualisation. Some related works are summarised in Table 6.9.

Recently, XML, which is regarded as the next-generation Internet mark-up language and has inherent relationships with some emerging Internet technologies, such as Web services, has received more attention as the exchange format for design and manufacturing data in a collaborative environment. XML format offers several potential advantages and allows for tag definition reflecting the tailored structure of the design and manufacturing data to facilitate data exchange between different functional modules. Some potentials of XML in collaborative design and manufacturing can be briefly summarised as follows:

- XML can define data as a document-oriented presentation format. To facilitate communication and exchange of information. XML separates the logical structure of a document from its presentation to define a document-oriented presentation, and a user can specify several styles for the same XML document. With such characteristics, the same document structure can be defined as several output formats for different applications in the design and manufacture domains.
- XML can build documents from heterogeneous data and information. An integrated collaborative design and manufacturing application usually includes heterogeneous information sources, such as documents, knowledge bases, relational or object-oriented databases, case bases, etc. XML enables the data and information from several sources to be mixed. For instance, data might come from a geometrical database, while text might come from a document management environment.

- XML can provide multiple views of the same data. This characteristic enables the same information to be processed differently. The presentation of different views of the same information could greatly assist user-friendly diffusion of the information among different classes of applications in design and manufacturing.

- In order to leverage the semantic power of the EXPRESS language in STEP, there is a desire to develop a combined strategy of STEP and XML. The ISO 10303-28 (XML representation of EXPRESS schemas and data) launched in the late 2000 represents a valuable step in this direction. However, the large size of XML files converted from EXPRESS causes a low-efficiency in the data transferring for the limitation of bandwidth. Another drawback is that an XML file from EXPRESS lacks flexibility. These issues should be considered carefully in the future work [Burkett, 2001; STEP Tools Inc.]

(2) Service-centric integration

Several significant service-centric mechanisms for collaborative system integration have been reported, including interface-wrapping services, agent-based services and Web services. To address the inherent complexity of the structures and interactions among the functional modules in an integrated system, a common "component interfaces" mechanism has been developed, through which various application components can be made available as services to interact with each other. Towards this direction, several competing technologies, such as DCOM (Distributed Component Object Model) by Microsoft Inc., CORBA by the Object Management Group (OMG), and J2EE (Java 2 Platform, Enterprise Edition) by Sun Inc., have been launched. These technologies are comparable and the properties of these technologies lend themselves to different situations.

Table 6.9 Related works for data-centric collaboration.

R&D Works	Functional Characteristics	Communications and Infrastructures
Chen and Liang, 2000	A system integrating and sharing engineering information to support CE activities such as domain investigation, functional requirement analysis, and system design and modelling.	• Exchange data: STEP AP203, VRML • CORBA (Visi Broker)-based web client/server • VRML plug-in web browser
Zhao, *et al.*, 2000	A system for product information exchange and sharing among distributed CAD/CAM users with different platforms.	• Exchange data: proprietary CAD files • Two-tier client/server • TCP/IP socket communication
Cheng, *et al.*, 2001	A Web-based design and manufacturing support system, including seven functional modules: electronic catalogue, intelligent selection, mounting details, sealing devices, lubrication, manufacturing database, and design module.	• Exchange data: CAD drawing files • Three-tier client/server based on Java RMI and JDBC
Kong, *et al.*, 2002	An Internet-base collaborative system for a press-die design process for automobile manufacturers.	• Exchange data: CAD files • Three-tier client/server based on CORBA, Java, Java3D and relational database
Zhou, *et al.*, 2003	An Internet-based system for designers to look for and retrieve distributive design knowledge, which is represented according to STEP standards and an ANN is used for knowledge search engine.	• Exchange data: STEP files • Three-tier client/server based on ASP and OCBC

Liu [2000] proposed a Microsoft COM interface-based framework to wrap and expose API functions of CAD kernels/systems and process planning modules for remote invocations. The concept of developing standard interface specifications, namely the common core interfaces, was proposed to encapsulate specific feature functions of different CAD kernels/systems to provide a generic and neutral application layer according to some international standards for features. The advantages of the work include the straight-forwardness of calling wrapped feature functions and the neutrality of CAD/CAPP/CAM kernels/systems for different applications. Chao, *et al.* [2002] proposed a CORBA-based framework as a wrapper to facilitate the communication between design and manufacturing tools. Within the CORBA-based framework, an Interface Definition Language (IDL) is used to define the interfaces of the functional modules for services to realise their interoperation. In the Web-based fixture design and manufacturing system presented by Mervyn, *et al.* [2004], Java/RMI (one of the J2EE technologies) is used to wrap the remote fixture design and manufacturing methods as services for remote invocation and manipulation. To alleviate the complexity of the system caused by the add-on wrapping structures and the huge programming effort for implementation, Li, *et al.* [2005(b)] proposed a multi-layer wrapping mechanism based on Java Servlet (one of the J2EE technologies) to extract some common wrapping parts from different functional modules to form an independent middle layer. This middle layer serves as a communication channel between the wrapped functional modules and other services in the Internet environment, and the load of each functional service can be reduced.

The aforementioned interface-wrapping mechanism set-up a foundation to build up other advanced mechanisms, such as Multi-Agent Systems (MASs) and Web services. A MAS, which is a self-adaptive distributed artificial intelligent technology to organise simple systems into a complex one, provides an advanced service integration solution to leverage the development of collaborative systems. Existing and legacy software systems can be encapsulated as agents to integrate design, planning, scheduling, simulation, execution, and product distribution, with those of their suppliers, customers and partners to collaborate in an open, distributed intelligent environment via networks. A Web service, which is based on XML schemas and a communication protocol SOAP (Simple Object Access Protocol) to provide a neutral data exchange format, can be published, located, and invoked across the Web. Once a Web service is deployed, other applications (and other Web services) can discover and invoke the deployed service. Some features of MASs and Web services are summarised as follows:

- The developments of MASs and Web services are towards more standards to define services and their communication. For instance, MASs specify a KQML (Knowledge Query and Manipulation) language for communication, and Web services use XML to realise more generic communication.

- The interface-wrapping mechanism is only for developing and deploying services in the network, while in MASs and Web services, coordination mechanisms are one of the essential components to support collaborative activities. From this aspect, they are facing the same or similar research challenges with those of collaborative CAD and development systems mentioned earlier, including system architectures, communication and cooperation, dynamic system reconfiguration, knowledge representation, conflict resolution and management, distributed dynamic scheduling, etc.

- Meanwhile, from the research viewpoint, currently, there are different focuses for these two technologies. For instance, the research of MASs is to leverage on artificial intelligent technologies to make them more autonomous, robust and mobile for the system design, and smarter to learn from the past experience and surrounding environment. The issues for Web services include interoperability, cross-platform capabilities and information security. These research issues can complement each other or even be integrated to form a more advanced solution eventually, such as smart Web services.

Some systems based on MASs are summarised in Table 6.10, and some third-party MASs are summarised in Table 6.11.

Table 6.10 Some agent-based systems for collaborative design and manufacturing.

R&D Works	Functional characteristics	System characteristics
Jacquel and Salmon, 2000	Features from an ACIS modelling kernel are wrapped as services for remote design and manufacturing analysis.	The agent engine used her is the Swarm engine developed at Santa Fe Institute.
Shen, *et al.*, 2000	A MetaMorph agent architecture ensuring the coordination among design parts and resource agents to support distributed design and manufacturing activities.	The architecture is a mediator-centric hybrid agent organization. AutoCAD with AME 2.0 is used to support product design. TCP/IP protocol is used to support high-level KQML communication among agents.
Gerhard, *et al.*, 2001	An event-based and agential framework to communicate design and manufacturing information through agent channels, and manufacturing analysis functions are enveloped as agents to support the establishment of an open and plug-in environment.	Java RMI technology is used to establish the agent infrastructure. Exchanged information is wrapped as events for communications in the environment..
Sun, *et al*, 2001	An agent architecture integrating design, manufacturability analysis, process planning and scheduling.	The multi-agent organization is based on JATLite multi-agent system. TCP/IP protocol is used to support high-level KQML communication among agents.
Kotak, *et al*., 2003	A framework for holonic design and operations based on three functional modules – holonic control, virtual simulation and human/system integration.	The multi-agent organization is based on JADE platform, which is an open-source Java agent development framework.

Table 6.11 Some Java-based MASs and their features.

MASs	Features
ABLE (Agent Building and Learning Environment) by IBM www.research.ibm.com/able	ABLE is a Java-based framework for developing and deploying hybrid intelligent agents and applications. A set of reusable JavaBean components and several flexible interconnection methods are provided to support the combination of components as agents. Data access, filtering and transformation, learning and reasoning capabilities are implemented in ABLE.
Aglets by IBM www.trl.ibm.com/aglets	Aglets are Java-based autonomous agents to provide the basic capabilities of mobility. Aglets can communicate by using a whiteboard to enable agents to collaborate and share information asynchronously. Synchronous and asynchronous message passing is also supported. No inherent intelligence is supported at this moment.
JADE (JAva DEvelopment) by CSELT S.P.A., Italy Sharon.cselt.it/projects/jade	JADE provides a set of agent services such as agent-naming service, yellow page service, transport protocols and interaction protocols. The communication infrastructure enables agents to access private message queues. It supports Java/RMI and CORBA. It also supports complex scheduling of agent tasks as well as integration with the Java Expert System Shell (JESS) reasoning engine.
JATLite (Java Agent Template Lite) by Stanford University java.stanford.edu	JATLite is a set of light-weight Java packages based on a layered architecture. It supports typed-message and autonomous agents. Its development focus is on communication to provide a different communication protocol at each layer.
ZEUS by British Telecom www.labs.bt.com/prjects/agents.htm	ZEUS is a framework for the development of collaborative agent systems that are constructed by using large, coarse-grained agents to emphasise autonomy and cooperation. The agents cooperate to combine their individual limited resources and knowledge to solve large problems.

6.5 Summary

Research and development have been actively carried out to develop prototype systems and methodologies to support collaborative CAD and development systems. In the recent years, CAD and software vendors have quickly realised the huge business opportunities in this area and launched a number of commercial systems in the markets. In this chapter, according to the functions and roles of the users participating in a design activity, some relevant research and systems are summarised in three threads: visualisation-based systems, co-design systems (these two for horizontal collaboration) and hierarchical collaborative systems. The focuses of horizontal collaboration are on the collocation of a team from the upstream design stage to carry out a complex design task synchronously or asynchronously, while the focuses for hierarchical collaboration are to establish an effective communication channel between the upstream design and the downstream manufacturing to enrich principles and methodologies of concurrent engineering to link diversified engineering tools dynamically. In each category, distributed architectures, information representation and coordination mechanisms are discussed in detail.

The future trends for collaborative systems include:

- Integration of horizontal and hierarchical collaboration. Horizontal and hierarchical collaborations are complementary in functions. It is important to establish a vertical seamless linkage between the upstream design and the downstream manufacturing processes through the creation of intelligent strategies for effective information exchange, and a horizontal interpersonal linkage of the group work in the upstream design phases. This integrated system can support interrelated activities and share domain knowledge between designers and systems to improve design quality and efficiency. Modules for the hierarchical collaboration should be wrapped as services for remote revoking. Within this system, scheduling and coordination is becoming more crucial and challenging, and distributed intelligent algorithms and technologies, such as MASs or Web services, can be used to enhance the integrated system.
- Security and Interoperability of collaborative systems. As customers, suppliers and designers from different places move to Internet-based collaboration, security must be considered carefully. While much of the security solutions offered by the current collaborative systems will be handled at the transmission level, they can also benefit by incorporating additional security features into their data models. Enhanced interoperability between collaborative design systems, and collaborative design and PDM systems, need to be achieved. IGES

and STEP are currently the de facto standards for small and medium enterprises and suppliers. Therefore, at a minimum, collaborative design solutions must be able to successfully handle IGES and STEP importing and exporting between the major collaborative applications to realise data-centric integration and interoperability. Ultimately, the goal for collaborative solutions must include the ability to access and manipulates legacy design and CAD data in their native file formats for various services and applications.

- More advanced feature- and assembly-based methodologies in collaborative systems for efficient sharing of information and multiple domain applications. The current collaborative systems are not generally accepted in industry. Besides the reason that different cultures, educational backgrounds, or design habits of designers hinder an effective collaboration, another major problem for the systems is due to the weakness in interactive capabilities, real-time and convenient collaboration. Effective distribution/collaboration algorithms are imperative to develop new feature- and assembly-based methodologies to improve the communication and cooperation efficiency. Some promising directions for developing the methodologies and algorithms include simplifying design models to eliminate some unnecessary exchanged information of the models when co-design to reduce the width requirements of the Internet, and incrementally transmitting models for streaming (frame-by-frame) communication. Meanwhile, in order to support feature-based applications to cross domains between design and manufacturing to support distributed hierarchical collaboration, algorithms need to be explored to realise bi-directional communications and information conversions between various application domains.

Chapter 7

Development of a Web-based Process Planning Optimisation System

The Web is a popularly Internet tool used to provide a light-weight and an operation system-independent platform for users to search, browse, retrieve and manipulate information disseminated and shared remotely. A Web-based application system can facilitate an on-line team to take on design discussion, product review, design remark and customer survey to enhance collaborative new products and conceptual design. In this chapter, a Web-based prototype system has been set-up for users to carry out visualisation-based manipulations and process planning of design models to support distributed design and manufacturing analysis. The process planning module, which has been deployed as a service in the Internet, consists of four intelligent methods. This Web-based system has been integrated with a distributed feature-based design system. Through effective utilisation of the Web and the Java technologies, this system is independent of the operating system, scalable and service-oriented, and can be used by a geographically distributed design team to organise concurrent engineering design activities effectively.

7.1 Introduction

Based on the Web, design plans can be dynamically shared and updated in an Internet environment and conveniently accessed and manipulated by remotely located people in the design team, management, marketing, maintenance and customers for efficient design collaboration, design process monitoring, product pre-review and evaluation. Application services in product design, process planning, engineering analysis and simulation, can be conveniently embedded in a Web environment as ASPs for remote invoking and manipulation. Realising the merits of the Web technology, researchers and developers have been actively exploring and developing Web-based design and manufacturing systems, and the work can be summarised from the following two aspects (more details in Chapter 6).

(1) Web-based visualisation systems

Web-based visualisation systems [Cimmetry Systems Autovue™ (www.cimmetry.com), Actify SpinFire™ (www.actify.com), Autodesk Streamline™ (www.autodesk.com); etc.] have been developed to support visualisation, annotation and inspection of design models for distributed design and manufacturing activities. These visualisation systems are light-weight, easily-deployed and platform-independent, and can facilitate an on-line team to take on design review, discussion, remark, and customer survey to enhance collaborative product design and analysis. To deliver and manipulate interactive 3D objects effectively on the Web, some concise formats specially designed for Web applications, such as VRML, X3D, W3D (Web 3D) and MPEG-4, have been launched to represent the geometry of 3D models as visualisation-used triangular meshes, trimming lines and some attributes [Web3D; Roy and Kodkani, 1999; Kan, *et al.*, 2001; Huang, 2002; Zhang, *et al.*, 2004]. VRML is a fundamental for the series of standards to represent geometric elements and scenes, whilst X3D and MPEG-4 are extended to support representation based on XML and video/audio application in compressed binary formats, respectively. Some formats, such as OpenHSF [Hoops™ (www.openhsf.org)] and XGL/ZGL [Autodesk Streamline™ (www.autodesk.com)] are functionally equivalent to VRML in geometric representation whilst they define data and algorithms for effective 3D streaming transmission over the Internet through data compression, mesh simplification and object prioritising.

The Web-based systems are based on the HTTP communication protocol, and there are two basic types of programming, i.e., server-side and client-side programming [Roy and Kodkani, 1999; Huang, 2002; Kan, *et al.*, 2001; Zhang, *et al.*, 2004]. For instance, Java Servlet,

Microsoft ASP* and CGI, which belong to the first type, can execute codes at the server side to generate display information at a Web browser (client). Java Applet and Microsoft ActiveX, which are examples of the second type, download codes from the server and execute them at a Web browser. Considering the large-volume 3D data and requirements for frequent interactive operations, hybrid architectures are mostly designed in which the client-side programming is used to support the establishment of the visualisation systems embedded in a Web browser and the server-side programming is used to maintain the information communication between the clients and a server. As Java APIs for light-weight 2D and 3D models (such as VRML models), Java2D and 3D have been popularly used in a Web browser for graphics rendering and manipulations.

(2) Web-based design and manufacturing systems

The researches reported in this category have different characteristics and implementation strategies due to their diversified functions and applications. However, from the system infrastructures and information communication mechanisms perspectives, they share some similar development features and trends.

Chen and Liang [2000] proposed a Web-based system to integrate and share engineering information to support design and manufacturing activities, such as domain investigation, functional requirement analysis, and system design and modelling. Functional modules in their system are wrapped and supported by CORBA for communication. The CyberCut system developed at the University of California at Berkeley [Sung, *et al.*, 2001] is a Web-based system integrating product design and process planning as a Java Applet program, and includes three primary modules, namely, (1) a Web-based feature-based design tool to model prismatic 2.5D parts using a Destructive Solid Geometry (DSG) approach; (2) a new geometric representation based on the DSG for information

---

* The Microsoft ASP is different from the Application Service Providers (ASPs) mentioned earlier. The former is a programming technology and the latter is a new application and business model. The ASP(s) in the following part of the chapter refers to the latter unless specified.

exchange between the design and process planning modules; and (3) an automated process planning system consisting of macro planning and micro planning. The FIPER (Federated Intelligent Product EnviRonment) system [FIPER Project (www.fiperproject.com/fiperindex.htm); Bailey and VerDuin, 2000; Beiter and Ishii, 2003] funded by the National Institute of Standard and Technology developed a new product design and analysis technology. The main objective of this system is to develop a Web-based distributed framework for design analysis and product lifecycle support based on component mechanisms and configurable workflow mechanisms. It can provide open and flexible capabilities to incorporate existing analysis and design tools/methods through Java-based wrapping mechanisms, which include the Java Native Interface (JNI) and the FIPER SDK toolkit. A workflow for a design process can be conveniently organised and configured by the users through assembling components in the distributed environment. Xiao, *et al.* [2001] developed the Web-DPR system as an infrastructure to support collaborative design and manufacturing. Based on the Java RMI mechanism, agents and an event-based mechanism, the functional modules of the systems can be linked and coordinated effectively. Shyamsundar and Gadh [2002] developed a new geometric representation named as AREP and a collaborative prototyping system based on this representation to perform real-time geometric modification for components/sub-assemblies in an assembly model. Mervyn, *et al.* [2003] proposed a Web-based fixture design system, in which an XML format was designed for the transfer of information and knowledge between functional modules in the distributed environment. In the work of Choi, *et al.* [2003], Web service architectures were utilised to establish a new generation of distributed design and manufacturing platform based on the XML schemas and a communication protocol SOAP to provide a neutral data exchange format and effective capabilities in interoperability, integration and Internet accessibility of services.

Based on the technical characteristics of these works, the following research issues reflect the current development features and trends:

- The concepts of services and ASPs are becoming more popular in Web applications. The installation of application modules in remote servers as services and execute them from the Web browsers offer many advantages, such as avoiding complicated installations for individual computers, easy upgrading of application modules and lowering the acquisition costs for small and medium enterprise through renting these services. One of the crucial research issues is to propose an adaptable wrapping mechanism for application systems to be used as services on the Web.

• Most of the current design systems are facilitated to export a proprietary model as a VRML model. However, during the conversion process, the organisation and relevant property information based on features, which are relatively important in CAD, CAPP and CAM systems for the high-level aggregations and interpretations of geometric and topological information, is lost. Due to this limitation, most of the Web-based visualisation systems cannot effectively support certain on-line manipulation operations on features, such as selecting a feature or its properties for manipulation, highlighting or hiding a feature, or dynamically attaching attributes to a feature in a design model for evaluation and analysis. Meanwhile, XML has been actively explored to support complex engineering information to take advantage of its design features and effective functions in the Internet applications. Hence, it is imperative to develop an XML-based format that can represent engineering data based on features, has strong interoperability and cross-platform capabilities and is suitable for Web applications.

• To facilitate product design and realisation processes on the Web, a visualisation-based system and the remote application services need to be integrated. The former can provide a Web-based environment for users to retrieve, view and manipulate a design model for design review, analysis or simulation conveniently. The latter deployed in the Internet can be invoked by users dynamically through the Web to evaluate and optimise the design so as to implement a collaborative concurrent engineering methodology.

In this chapter, a Web-based distributed design and manufacturing system has been developed to serve as a platform for distributed users to carry out process planning activities through optimising the selection of machining resources, determination of set-up plans and sequencing of machining operations of a design model. Four intelligent methods have been developed to solve this optimisation problem. The optimisation module is deployed in the Internet as a Java Servlet-based application service based on a multiple-layer wrapping mechanism. Java Applet, Java2D and 3D technologies have been used in the system to develop a visualisation-based manipulation environment for the design models and the optimisation results. Through an XML-based data exchange format based on features and VRML, this system can exchange information with a distributed feature-based design system to form an integrated design and manufacturing analysis environment across the Internet.

7.2 A Tabu Search-based Optimisation Method

In Chapter 5, a hybrid GA/SA method has been developed to solve the process planning optimisation problem. Here, another optimisation method based on a TS (Tabu Search) method has been designed and embedded into the Web-based system together with the hybrid GA/SA method, a single GA method and a single SA method.

7.2.1 *Methodology for process planning Optimisation*

As the crucial activities in a process planning system, selecting suitable set-up plans, determining machining resources, such as machines and cutters, and optimising the sequence of the machining operations are relatively important to ensure satisfactory solutions with lowest machining costs. Considering that decision processes for these aspects are sometimes contradicting, and the evaluation criteria with different consideration perspectives can also be conflicting in some cases, some developed reasoning methods, such as the knowledge-based reasoning [Chang, 1990; Wong and Siu, 1995; Chu and Gadh, 1996; Wu and Chang, 1998; Tseng and Liu, 2001], graph manipulation [Chen and LeClair, 1994; Irani *et al.*, 1995; Lin *et al.*, 1998], Petri-nets based approach [Kruth and Detand, 1992] and fuzzy logic reasoning [Zhang and Huang, 1994; Ong and Nee, 1994; Gu *et al.*, 1997], evolutional algorithm and heuristic reasoning-based approach [Vancza and Markus,

1991; Chen and LeClair, 1994; Yip-Hoi and Dutta, 1996; Zhang, *et al.*, 1997; Chen, *et al.*, 1998; Reddy, *et al.*, 1999; Qiao, *et al.*, 2000; Ma, *et al.*, 2000; Lee, *et al.*, 2001] cannot effectively solve this problem with a global optimised result. Recently, evolutional and heuristic algorithms have been applied to the process planning research, and multiple objectives, such as the minimum use of expensive machines and tools, minimum number of set-ups and tool changes, and achieving good manufacturing practice, have been incorporated and considered as a unified model to achieve a global optimal target [Zhang, *et al.*, 1997; Li, *et al.*, 2002]. Four approaches, including the GA, SA, hybrid GA/SA and TS, have been applied to solve this problem.

A part consists of a series of features, and each feature can be mapped to a set of machining operations which form the elements of a process plan. A process plan usually needs to determine the sequence of the operations, set-ups, and the machine and cutter for each operation. The objective is to generate the most economical plan evaluated from the numbers of required set-ups and the utilisation of machining resources. In each operation, there is a set of alternative machines, cutting tools and set-ups (represented as TADs) under which the operations can be executed. The proposed TS approach determines an optimised sequence of the operations and the utilised machine, cutting tool and TAD chosen from their corresponding candidates according to certain optimisation criteria, which include the costs for the utilised machines (TMC), utilised tools (TTC), Set-ups (TSC), machine changes ($TMCC$) and tool changes ($TTCC$). For violated constraints, a penalty cost (PC) is computed. The detailed formulas and computation processes for these criteria can be found in Chapter 5. The geometric and manufacturing interactions between features as well as the technological requirements in a part can cause some preliminary precedence constraints between the machining operations. These interactions and technological requirements can be summarised into several types: (1) fixture interactions, (2) tool interactions, (3) datum interaction, (4) thin-wall interactions, (5) feature priorities, (6) material-removal interactions, and (7) fixed order of machining operations. The definitions and examples of these constraints are given in Chapter 5.

Based on these concepts, the process planning optimisation model can be represented as follows:

$$\underset{x}{Min} \quad (Machining\_Cost(x) + Penalty\_Cost(Constra \text{ int } s))$$
$$x \in \{\text{Trial process plans}\}$$

In this model, *Machining _ Cost* is the total machining cost consisting of *TMC* , *TTC* , *TSC* , *TMCC* and *TTCC* . As some constraints are involved and the initial optimisation problem is constraint-based, a penalty cost for the violated constraints is considered to solve it as a non-constraint optimisation problem. The optimisation objective is to reduce the total machining cost. Some details for an improved TS-based approach, which can generate process plans with near-optimal or optimal results according to the optimisation objective, will be described next.

7.2.2 *Algorithm description*

A typical TS is used here and it consists of three main strategies, viz., the forbidding strategy, freeing strategy and aspiration strategy [Glover, 1997]. The forbidding strategy controls the solution that enters the Tabu list. The freeing strategy is used to manage the solution that exits the Tabu list and when it exits. The aspiration strategy is the interplay between the forbidding and freeing strategies for selecting trial solutions. The workflow of the algorithm is shown in Fig. 7.1. Some basic elements in the algorithm are briefly stated as follows.

(1) Initial plan, current plan and elite plan. An initial plan is generated with *n* operations. The sequence of the *n* operations is randomly arranged, and the machine, tool and TAD for the execution of an operation in the plan are randomly determined from the corresponding candidate lists. The current plan is the optimised solution in each iteration and used for generating the neighbourhood trial solutions. An elite plan records the best solution found so far. A current plan might not be an elite plan since the current plan is only the best move in its neighbourhood trials at a particular iteration. Thus, it might be worse than the elite plan recorded thus far.

(2) Neighbourhood strategies. A set of variant plans can be generated from a current plan for trials using neighbourhood strategies. The

neighbourhood strategies, which are illustrated in Fig. 7.2, include two basic manipulations. The first mutation manipulation randomly chooses a set of machines, tools and TADs from the alternative lists to replace the current ones in the operations of a plan. The second manipulation changes the sequence of two operations in a plan using shifting, swapping or adjacent swapping operations. The size of the variant plans and four probabilities of applying the mutation, shifting, swapping and adjacent swapping operations are represented as S_{TS_N}, P_{TS_mu}, P_{TS_sh}, P_{TS_sw} and P_{TS_as} respectively.

(3) Forbidding and freeing strategies. The forbidding strategy can avoid cycling and local minimums by forbidding certain moves during the most recent computational iterations. The freeing strategy is used to delete the forbidding restrictions of some solutions so that they can be reconsidered in the future search. A Tabu list, which is stored as a linked list and managed as a "first-in-first-out" (FIFO) queue, is employed to store recently visited current plans to realise these strategies. The size of the Tabu list, T_s, plays a crucial role in the search for high-quality solutions and should be critically analysed since a T_{SA_s} that is too small might cause a high occurrence of cycling, and a T_{SA_s} that is too large might forbid too many moves and deteriorate the solution quality. Using the *HD* concept defined in Equation (5.13), a current plan $Oper_i$ can be compared with a plan $Oper_j$ stored in the Tabu list to determine their similarity, and if and when to apply the forbidding or freeing strategies.

(4) Aspiration strategy. When a plan is forbidden by Tabu restrictions, an aspiration strategy can enable this plan to become acceptable as a current plan if it satisfies a certain criterion. This strategy can provide some flexibility to the Tabu restrictions by leading the move in a desirable direction. A common criterion is to override a tabooed plan if its machining cost is lower than that of the elite plan.

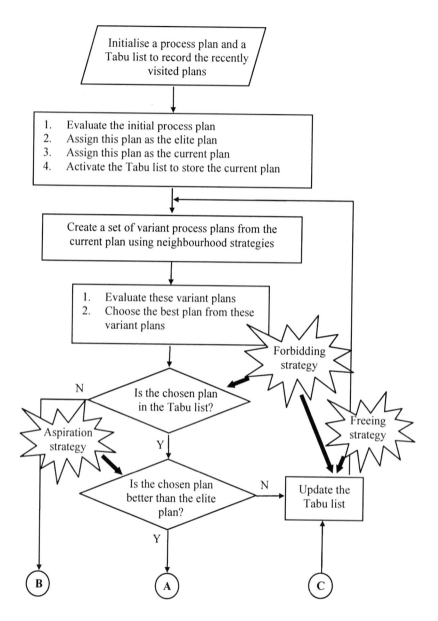

Fig. 7.1 The workflow of the TS process planning optimisation.

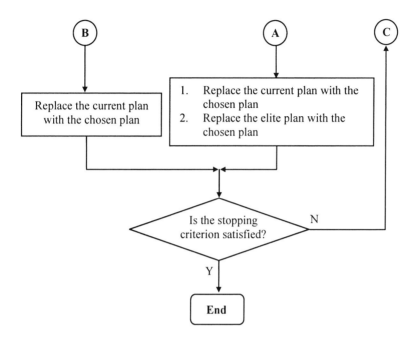

Fig. 7.1 The workflow of the TS process planning optimisation (cont'd).

(5) Stopping criteria. Termination conditions for the searching algorithm can be set using one of the following criteria: (a) the iterations reach a pre-defined number; (b) the computation is carried out continuously for a pre-defined number of iterations after the ratio between the machining cost of the generated elite plan and a pre-specified value (an optimal or near-optimal value determined using other methods such as a GA or SA) falls into a certain range (e.g., 1%); and (c) the elite plan is kept unchanged for a pre-defined number of iterations.

(6) Constraints handling method. In TS, the precedence constraints in a part should be considered to ensure the manufacturing feasibility of process plans. However, similar to the situation in the GA/SA process, the sequences of an initial plan or a variant plan generated using neighbourhood strategies might be inconsistent with these constraints. The constraints handling method proposed in Chapter 5 is used to handle the hard and soft constraints respectively.

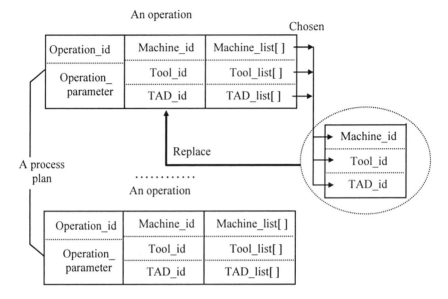

(a) *An operation is changed by mutating the determined machining resources from the candidate list*

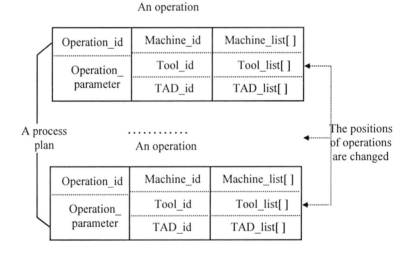

(b) *The sequence of operations is changed by shifting, swapping or adjacent swapping manipulations*

Fig. 7.2 Two basic manipulations to generate variant plans from a current plan.

7.2.3 Determination of parameters

The main parameters that determine the performance of the TS include the size of the Tabu list - T_{TS_s}, the size of the variant plans from a current plan - S_{TS_N}, and the four probabilities of applying the shifting, swapping, adjacent swapping and mutation operations - P_{TS_sh}, P_{TS_sw}, P_{as} and P_{TS_mu}.

T_{TS_s} plays an important role in the search for solutions. A small T_{TS_s} might cause a high occurrence of cycling, and a large T_{TS_s} might deteriorate the solution quality. Through trials, T_{TS_s} is chosen as 20 and the comparison results are illustrated in Fig. 7.3 (a) and (b). A suitable S_{TS_N} can ensure good computational efficiency and algorithm stability. Based on the observations from Fig. 7.3 (c) and (d), $S_{TS\_N} = 30$ is a good choice for the algorithm to achieve the objectives. Several groups of these four parameters are chosen to determine the suitable values. In Table 7.1, ten trials are conducted for each group of parameters, and the statistical mean, maximum and minimum machining costs for the ten final results are listed. It shows that the fourth group of parameters in the table ($P_{TS\_sh} = 0.85$, $P_{TS\_sw} = 0.85$, $P_{TS\_as} = 0.5$ and $P_{TS\_mu} = 0.85$) can yield good performance of the algorithm. These comparisons are based on Part 1 in Chapter 5. Based on the trials for Part 2 in Chapter 5 and using the same chosen parameters, satisfactory results were obtained. Hence, these parameters are generally applicable in other situations.

Table 7.1 Determination of four probabilities in the TS algorithm.

	P_{sh} (shifting), P_{sw} (swapping), P_{as} (adjacent swapping), P_{mu} (mutation)					
	1.0, 1.0, 1.0, 0.75	0.85, 1.0, 0.85, 1.0	0.5, 0.5, 0.5, 0.5	0.85, 0.85, 0.5, 0.85	0.5, 0.85, 0.85, 0.85	0.85, 0.85, 0.5, 0.85
Mean	1368.0	1366.0	1589.5	1340.5	1478.5	1419.0
Maximum	1463.0	1508.0	1758.0	1363.0	1918.0	1703.0
Minimum	1343.0	1343.0	1343.0	1328.0	1328.0	1328.0

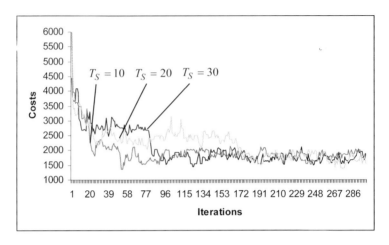

(a) *Machining costs of current plans under different* T_S

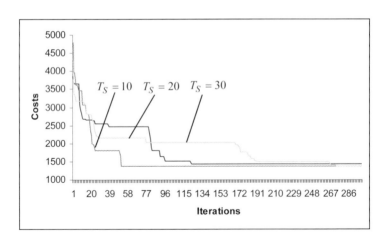

(b) *Machining costs of elite plans under different* T_S

Fig. 7.3 Determination of parameters of the TS algorithm.

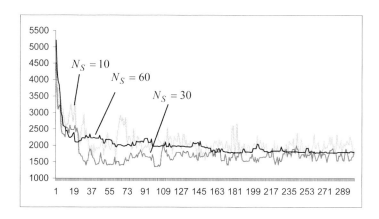

(c) *Machining costs of current plans under different* N_S

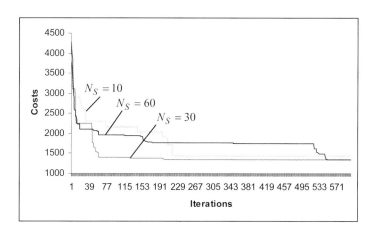

(d) *Machining costs of elite plans under different* N_S

Fig. 7.3 Determination of parameters of the TS algorithm (cont'd).

7.2.4 *Comparison studies of TS, SA and GA*

Part 1 and Part 2 in Chapter 5 are used to illustrate the performances of TS, SA and GA on the same optimisation model to give a comprehensive understanding of their characteristics. Conditions (1) and (2) are chosen for the studies for both Part 1 and Part 2. For Part 2, an additional Condition (3) of a dynamic workshop environment is tested.

(1) All machines and tools are available, and $w_1 - w_5$ in Equation (5.12) are set as 1.

(2) All machines and tools are available, and $w_2 = w_5 = 0, w_1 = w_3 = w_4 = 1$.

(3) Machine M_2 and Tool C_7 are down, $w_2 = w_5 = 0, w_1 = w_3 = w_4 = 1$.

The computations illustrated in Fig. 7.4 were performed for the two parts under Condition (1). The current plans of TS and SA at each iteration are used for generating the neighbourhood and next solutions, while in GA, each of them refers to the best plan chosen from a population in a generation (an iteration). Each of the elite plans is the best plan at each iteration of the three algorithms. Fig. 7.4 (a) and (b) are for Part 1, and (c) and (d) are for Part 2. It shows that the decreasing trends of the curves for TS and GA are smoother than that of SA. For SA, there are some "abrupt" decreasing points during its iteration process. Generally, TS and SA can achieve better (lower machining costs) solutions than GA, while TS can achieve a more stable performance than SA in terms of lower mean, maximum and minimum machining costs of the best process plans obtained. These observations are in accordance with the main characteristics of GA, which is prone to "premature" (converge too early and difficult to find the optimal or near-optimal solutions), and SA, which is vigilant to parameters and problems [Pham and Karaboga, 2000].

In Tables 7.2 and 7.3, more thorough comparisons for the three algorithms on Part 1 under two conditions, and on Part 2 under three conditions are made. The computations are based on ten trials for each algorithm under each condition respectively (each computation was conducted for 30s using a PIII-800 computer with 256M memory). Similar observations on the characteristics of the three algorithms can be obtained. Comparing the present approach to the optimal or near-optimal results obtained previously by the authors using the hybrid GA/SA method, the lowest machining costs under Conditions (1) and (2) are the same, while in Condition (3), a lower machining cost (2580.0) has been found using the current method (2590.0 in the hybrid GA/SA method).

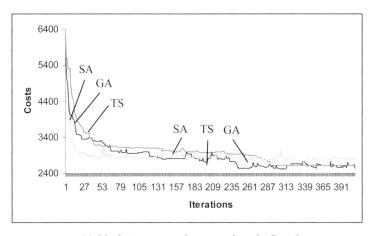

(a) *Machining costs of current plans for Part 1*

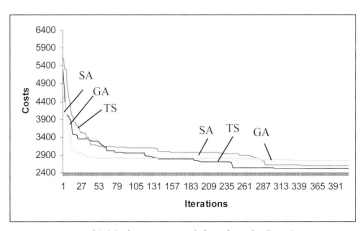

(b) *Machining costs of elite plans for Part 1*

Fig. 7.4 Comparison studies of three algorithms for the two parts.

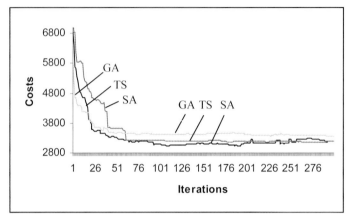

(c) *Machining costs of current plans for Part 2*

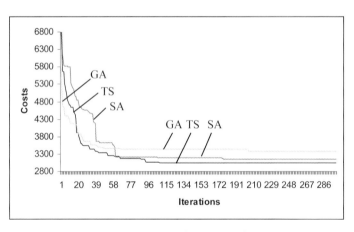

(d) *Machining costs of elite plans for Part 2*

Fig. 7.4 Comparison studies of three algorithms for the two parts (cont'd).

Table 7.2 Comparison studies of three algorithms for Part 1 under two conditions.

		Tabu Search	Simulated Annealing	Genetic Algorithm
Condition (1)	Mean	1342.0	1373.5	1611.0
	Maximum	1378.0	1518.0	1778.0
	Minimum	1328.0	1328.0	1478.0
Condition (2)	Mean	1194.0	1217.0	1482.0
	Maximum	1290.0	1345.0	1650.0
	Minimum	1170.0	1170.0	1410.0

Table 7.3 Comparison studies of three algorithms for Part 2 under three conditions.

		Tabu Search	Simulated Annealing	Genetic Algorithm
Condition (1)	Mean	2609.6	2668.5	2796.0
	Maximum	2690.0	2829.0	2885.0
	Minimum	2527.0	2535.0	2667.0
Condition (2)	Mean	2208.0	2287.0	2370.0
	Maximum	2390.0	2380.0	2580.0
	Minimum	2120.0	2120.0	2220.0
Condition (3)	Mean	2630.0	2630.0	2705.0
	Maximum	2740.0	2740.0	2840.0
	Minimum	2580.0	2590.0	2600.0

7.3 Infrastructure Design of the Web-based System

7.3.1 *System architecture*

The Web-based system is based on a multiple-layer client/server architecture and consists of four functional modules, namely, (1) a front-end client embedded in a Web browser to support the visualisation of design models, invocation of remote process planning optimisation services, and display and manipulation of optimisation results, (2) a look-up service to register, manage and search for services deployed in the Internet, (3) the process planning services deployed in the Internet, and (4) a database system (MySQL™ (www.mysql.com)) for storing information of the available machines, cutters and their costs. This system can communicate with a distributed feature-based design system, which will be discussed in Chapter 8, to retrieve design models represented in an XML format. The scenario is shown in Fig. 7.5.

The front-end client has four basic functions, namely, (1) a visualisation environment for manipulating design models and their optimised process planning results, (2) an interface to register services that is distributed in the Internet for remote calling and operations, (3) an interface to the database to retrieve and store information, and (4) an interface to the distributed feature-based design system to exchange design models and relevant information for visualisation and analysis in this system. Based on the Java Applet, Servlet and Java2D and 3D technologies, the front-end client includes three components, namely, (1) a Tomcat Web server, (2) an Applet for visualisation-based manipulations of design models and optimisation results based on Java 3D and 2D technologies respectively, and (3) communication facilities in the Applet to exchange information with the process planning services, database and distributed feature-based design system.

The addresses of the process planning services located in the Internet are registered in a look-up service of the system for remote calling (the look-up service in Fig. 7.5). To provide an extensible and flexible mechanism to integrate the application programs, a three-layered architecture, including wrapper classes for the services, abstract classes for the services, and detailed class and method implementations, has

overlapping features can include a combination (two features combined through an explicit Boolean operation) and an attachment (a new feature added to an existing feature through a local operation).

- Face attributes, such as types of faces (planar faces or surfaces), normals of planar faces and curvatures of surfaces.
- VRML patches for each face. The information is organised into two groups, namely, a group for the coordinates of the vertices and a group for their indices.

7.4 A Case Study

A case study is discussed to illustrate the process planning optimisation process in the Web-based system.

(1) A model can be created in a feature-based design system (shown in Fig. 7.10). As a manufacturing feature recognition service has not yet been integrated with the Web-based system, the design model is constructed with subtractive features removed from the initial building stock (manufacturing features).

(2) The model can be converted into an XML-based format and passed to the Web-based system (shown in Fig. 7.11). A user can observe the design model in the Web-based system. He can change the visualisation mode of the part, e.g., hiding the meshes, highlighting a feature, retrieving its parameters, rotating and zooming the part, etc.

(3) He can dynamically select and invoke a process planning service from the Web-based system. He can select an optimisation objective and the relevant weights, the parameters of the optimisation service, a feature or a face for specifying constraints, and the machining resources, such as the cutting tools and machines. Some intermediate results are shown in Fig. 7.12. The detailed information and machining resource of this part can be found in Chapter 5 (Part 2). The final optimised process plan and the costs are shown in Fig. 7.13. Fig. 7.14 is a Java2D-based visualisation and manipulation window in the system to observe, query, zoom and edit the optimisation results.

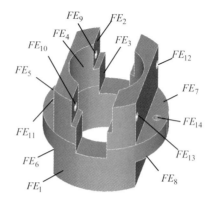

(a) *A design part with form features in B-Rep*

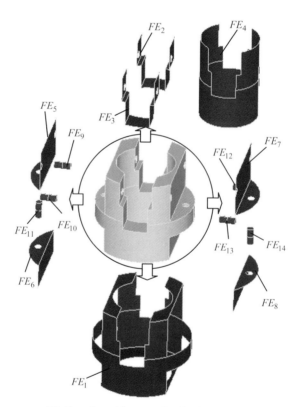

(b) *Face-based features for the design model*

Fig. 7.8 The conversion process from features in solids to features in meshes.

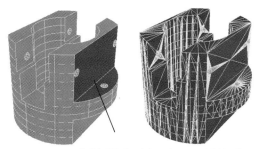

FE₇ is highlighted (meshes are hidden)

(c) *Facet-based features for the design model*

Fig. 7.8 The conversion process from features in solids to features in meshes (cont'd).

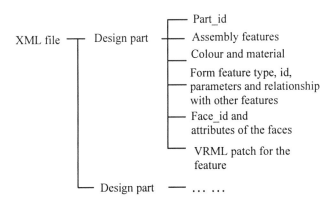

Fig. 7.9 An XML representation for a design model.

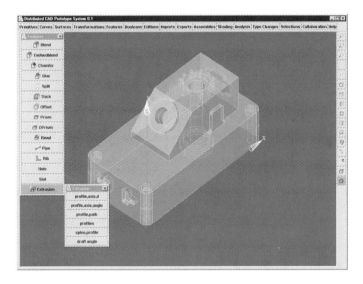

Fig. 7.10 A part is created in a feature-based system.

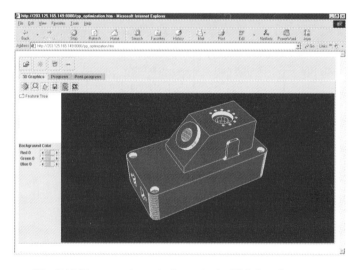

Fig. 7.11 The created part is shown in the Web-based system.

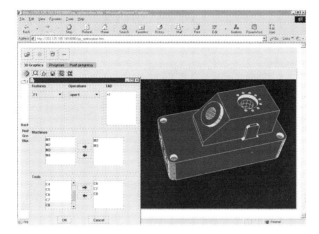

(a) *Selecting machining resources*

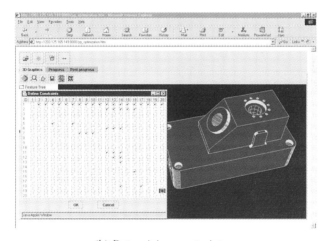

(b) *Determining constraints*

Fig. 7.12 Some intermediate processes for inputting the
optimisation information for the part.

(c) *Determining optimisation parameters*

Fig. 7.12 Some intermediate processes for inputting the
optimisation information for the part (cont'd).

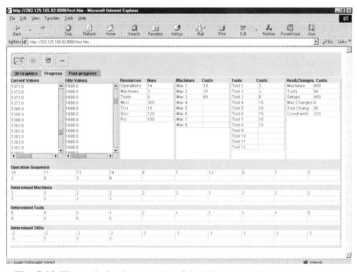

Fig. 7.13 The optimisation results of the TS service for the design part.

7.5 Summary

In this chapter, a Web-based process planning optimisation system to support distributed design is presented. This provides a convenient platform for users to view and evaluate a design model effectively through dynamically invoking remote process planning optimisation services. A TS-based approach is explained in detail to show the process. Through a wrapping mechanism, the optimisation module is equipped as services located in the Internet for remote invoking. A distributed feature-based design system can generate design models in an XML-style feature representation to provide for the Web-based system for viewing-based manipulations.

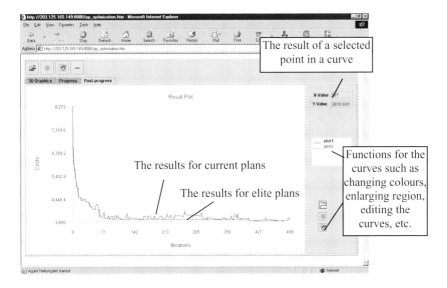

(a) *A window to manipulate the optimisation results of the design part*

Fig. 7.14 A window to show and manipulate optimisation results.

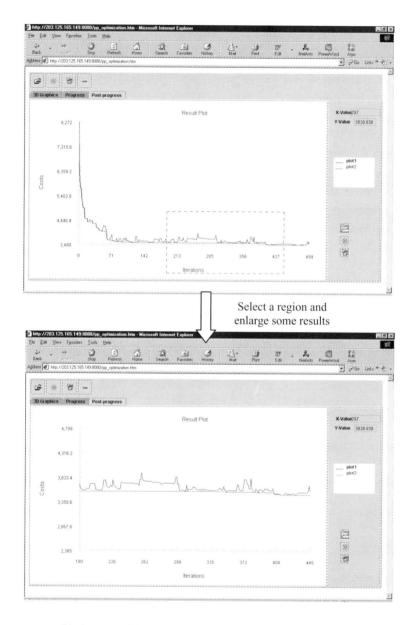

(b) *A region of the window is enlarged to highlight some results*

Fig. 7.14 A window to show and manipulate optimisation results (cont'd).

The main contributions of this work are summarised as in the following two aspects.

Firstly, through taking advantages of the effective utilisation of the Web and Java technologies, this system is independent of the operating system, scalable and service-oriented. The services located in the Internet can provide an effective manner for a designer to conduct a process planning optimisation process to evaluate a design by him or others in a distributed design activity.

Secondly, a new XML-style feature representation has been proposed to carry out some feature-based visualisation manipulations in the Web-based system. This format incorporates the characteristics of VRML and features to support Web applications. The XML-based information representation enables the system to be effectively adaptable to meet the new development of the Internet technology such as the emerging Web-service technology. The multiple-layer architecture designed for wrapping the process planning services allows the system with the services to be extensible to integrate other legacy systems.

Chapter 8

Distributed and Collaborative Design-by-Feature System

With the intense competition in the global marketplace, to develop a successful product requires geographically dispersed designers to build and test solutions collaboratively in a distributed infrastructure. Design-by-feature systems, which have been extensively used in product design and development, are imperative in providing distributive and collaboration-enabling support for global design collaboration. In this chapter, a prototype system (co-design system) has been developed to enable a dispersed team to accomplish a feature-based design task collaboratively. In the co-design system, one of the major challenges is to share design models and design changes among a working team efficiently, which is usually hindered by the large data volume. To address this issue, a distributed feature manipulation mechanism has been designed to filter the varied information of a working part during a co-design activity to avoid unnecessary re-transferring of the complete large-size CAD files each time when any interactive operation is imposed on the model by a client.

8.1 Introduction

Typically, to carry out a complex design task, a large design team will be engaged, and the communication and collaboration among members in the team are crucial to enable design to be carried out effectively. Traditional product design is geographically limited and CAD systems are standalone, such that a design activity is usually organised within an enterprise. Presently, a significant trend for

281

manufacturing enterprises is to out-source collaborative product development to external designers and suppliers. With the rapid development of IT, future CAD systems are moving towards supporting distributed and collaborative design, in which geographically dispersed systems can be integrated and a virtual design team can be set up within an Internet/Intranet environment. Large firms, such as Boeing, Ford, Kodak, MacNeal-Schwendler and Structural Dynamics Research, have realised the potential of distributed design systems and formed an Enterprise-wide Engineering Consortium to optimise their business processes through studying and applying such kind of tools. As an industrial application example, Ford worked collaboratively with its acquired Volvo from their separate sites on car design based on a distributed design platform. Due to the emerging distributed CAD technology, small and medium enterprises with specific domain knowledge will be able to participate and co-operate in a design process with large firms. Renting on-line high-end CAD systems running on CATIA and MicroStation systems have now become affordable for small and medium enterprises or even individuals, and not just for the industry giants.

Recently, some work has been carried out to develop distributed CAD systems and methodologies. However, there are some practical issues that have not been addressed explicitly and satisfactorily. Some concerns, such as expensive IT infrastructure, difficulties in the maintenance of data consistency in a multiple-user environment, and intolerable long waiting-time for updating large-size CAD models across networks, hinder the popularity of using the new distributed CAD systems.

To address these problems, a distributed feature-based design environment has been established in this research based on a Java client/server architecture and an open-source solid modelling kernel – Open CASCADE™. The objectives of this work are: (a) to establish a virtual environment to facilitate distributed and collaborative feature-based design activities; and (b) to optimise the organisation and transmission of information based on feature representation and manipulation.

8.2 Distributed Feature-based Representation and Manipulations

8.2.1 *System framework*

The system framework is illustrated in Fig. 8.1 and briefly described in Table 8.1. The main components in this system are as follows:
- Design clients;
- A collaborative server;
- Downstream manufacturing modules;
- Event-driven communications; and
- Remote interfaces of methods and objects exposed for remote calling in the environment.

The system framework consists of two primary aspects, namely, (1) distributed feature representations and manipulations; and (2) communication and collaborative mechanisms. Their details are explained in the following sessions.

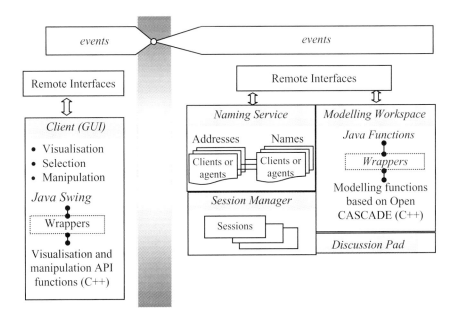

Fig. 8.1 System framework of the distributed design environment.

Table 8.1 Functional modules in the distributed and collaborative environment.

Clients	Collaborative Server
Parameter input/edition GUI	*Naming service*
• Inputting and editing parameters for features • Selecting geometric entities in features • Querying information for features and parts • Java JFC-swing components to set-up the GUI	• Registering the addresses (references) of clients and analysis services • Binding easily recognised names to the addresses of clients and analysis services for convenient invocation and manipulation
	Session manager
	• Dynamically generating working sessions • Sharing and updating a working part for clients registered in a working session
	Modelling workspace
Visualisation environment • Through Java Native Interface (JNI), some native API function for visualisation in the Open CASCADE can be invoked to display parts	• Accepting parameters or entities from the clients • Modelling parts based on Open CasCade™ solid kernel • Invoking and manipulating native API functions of the Open CASCADE™ through JNI
	Discussion community
	• Providing e-room for on-line discussion and information (text and multimedia) exchange among clients
Communications of information and invocations of remote methods	
• Communications in the environment are through an event-driven mechanism based on the Java RMI • A set of remote interfaces is declared for a set of methods to be invoked remotely based on the Java RMI	

8.2.2 *Distributed feature representations*

Based on an open-source solid modelling kernel, i.e., Open CASCADE™ (www.opencascade.com), a distributed feature-based modelling framework, which is organised as a "manipulation client + modelling server" scenario, has been developed to facilitate central provision and maintenance of information and services in a multiple-designer environment.

In this scenario, a client dynamically and interactively edits a part through inputting or adjusting the parameters of the features, and the server is responsible for modelling the part. Two representations, "light" on the client side and "heavy" on the server side respectively, have been proposed to fulfil the functional requirements and enhance the performance of the system.

A concise face-based representation, which depends on the visualisation modules that are separated from the Open CASCADE™, is established on the client side to support the interactive visualisation and manipulation functions (selection, transformation and changing visualisation properties of displayed parts). On the server side, a "heavy" representation with features and part information is set up and maintained to provide primary feature-based modelling functions.

On the server side, two data structures are established to store and maintain the information of a part and its features during the modelling processes, including Part Constructive Tree (*PC_Tree*) and Feature-to-Feature Graph (*F2F_Graph*). The *PC_Tree* is the primary data structure, which can generate the *F2F_Graph*.

A *PC_Tree* is used to organise the design features of a part in a hierarchical structure according to its evolving process. The data structure and main components, which are illustrated in Fig. 8.2, are as follows:

(1) The root of the tree is the constructive base feature of a part. Each intermediate node in the tree is a Boolean union or difference operator, and its corresponding leaf is an additive feature volumetrically added onto the part, or a subtractive feature volumetrically removed from the part. An auxiliary feature is attached to the root or a leaf as its auxiliary part.

(2) Each feature, which is associated with one or more datum planes or axes, surface tolerances, attributes, dimensions, etc., is represented as its initial shape in B-Rep. In these features, some faces or some portions of the faces are removed after Boolean operations. The remaining complete or partial faces, which are denoted as *revealed faces*, are used to compose a face-based representation for each feature and attached to the *PC_Tree*.

(3) During the evolving process of a part, the part is updated after each Boolean operation associated with the Boolean operations in the feature tree has been carried out.

(4) Features and faces are allocated different physical addresses in the distributed environment. An IDs mechanism for the features and the faces is designed on the server side to maintain the consistency of the entities in the distributed environment.

The relationships between features can be categorised as interacting and non-interacting relationships (the interacting relationships have been defined in Chapter 4). Based on these relationships, an *F2F_Graph* can be set up. On the client side, a feature tree with a structure similar to a *PC_Tree* on the server side is mapped, and each feature is represented as a *revealed face*-based compound to support visualisation and manipulation.

8.2.3 *Feature manipulations*

The modelling of a part is an incremental process, and the editions of a part include adding features, removing features, or modifying the parameters of features. The creation or edition of a feature in a part is through a bi-directional process in the distributed environment. Parameters of a feature are wrapped as a parameter event on the client side and transmitted to the server side for modelling. Object events are generated to organise *revealed face*-based compounds as features and feedback to the client side for visualisation. Some features are associated with other features in the part, and the modelling and editing operations on them might cause variations in the associated features. In order to speed up the transmission of the part efficiently via the network with a limited bandwidth, the varied information during the editing operation should be differentiated on the server side and synchronised with the unchanged information on the client side. A set of feature manipulation operations has been developed to support two processes – differentiation of features and differentiation of faces in features as four types: *Added*, *Deleted*, *Updated* and *Unchanged*. The scenario is shown in Fig. 8.3.

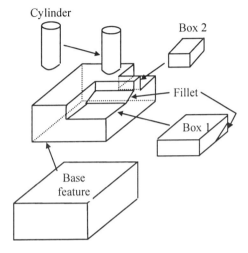

(a) *A part with five features*

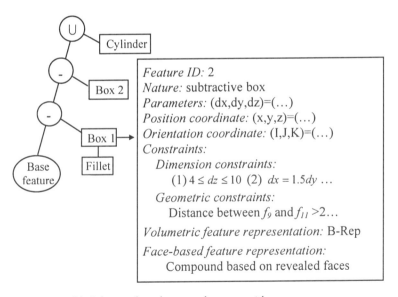

(b) *A feature-based tree on the server side*

Fig 8.2 Data structures of a part and its features on the server.

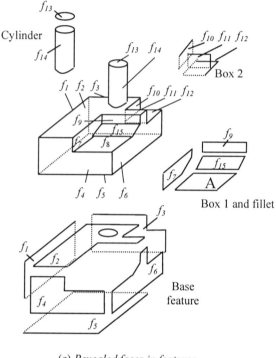

(c) *Revealed faces in features*

Fig. 8.2 Data structures of a part and its features on the server (cont'd).

The differentiation information is wrapped as events on the server side and broadcast to the clients. The classification processes, which are illustrated in Fig. 8.4, are described as follows:

(1) When a feature is added, this feature is *Added*. Its interacting features in the *F2F_Graph* are *Updated*. The features in the non-interacting set are *Unchanged*.

(2) When a feature is removed, this feature and the features nesting on this feature are *Deleted*. Features in which this feature nests on, and features with adjacent or overlapping relationships with this feature are *Updated*. Features in the constraining or non-interacting sets of this feature are *Unchanged*.

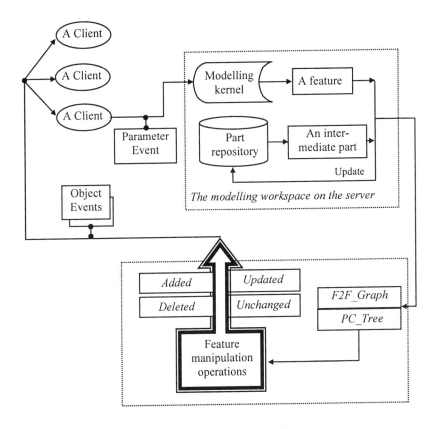

Fig. 8.3 The working processes for modelling a part.

(3) When the parameters of a feature are modified, the feature is *Updated*, and the process for other features consists of three steps:
  (a) Features, which are from the non-interacting set of this feature before the modification, that have adjacency, overlapping or nesting relationships (the interacting relationships between features can be found in Chapter 4) with this feature after the modifications are *Updated*;
  (b) Features, which are in the interacting set of this feature before the modification, will be differentiated into two types after the modification: features with changes in their *revealed faces* are

Updated, and features without changes in their *revealed faces* are *Unchanged*;

(c) If this feature has a constraining feature, the interacting and non-interacting sets of the constraining feature will be adjusted recursively according to the above steps.

(4) For features that are differentiated as *Added* from the above steps, the IDs of the features, the IDs and objects of the *revealed faces* are wrapped into object events for the features. For features that are *Deleted* or *Unchanged*, only their IDs are recorded in the events for the clients to erase or keep the relevant features and *revealed faces* information. For features that are *Updated*, their *revealed faces* are classified into *Added*, *Deleted*, *Updated*, or *Unchanged* as well. The data structure of the faces in an object event for an *Updated* feature includes two types: (1) for faces that are *Added* and *Updated*, the IDs of these faces and faces objects are filled into the object events; and (2) for faces that are *Deleted* and *Unchanged*, only their IDs are filled into the events for information deletion or retention.

These feature manipulation operations are illustrated by the examples in Fig. 8.5-8.7. In Fig. 8.5(a) and (b), a new feature, FE, is added to an intermediate part with eight features, and FE nests on an existing feature FE_1. The features in the part are differentiated as *Added*, *Updated* and *Unchanged*, as shown in Fig. 8.5(c). In Fig. 8.5(d), the process of differentiating the faces of the *Updated* feature, FE_1, is shown.

In Fig. 8.6(a) and (b), the features of a part and the *F2F_Graph* of these features are shown. In Fig. 8.6(c), for the deletion of FE_4 that nests on FE_1 and has non-interacting relationships with other features, FE_4 is *Deleted* and FE_1 is *Updated* while other features are *Unchanged*. The faces of the *Updated* FE_1 are classified further as *Updated* and *Unchanged* in Fig. 8.6(d). In Fig. 8.6(e), for the deleted FE_1, the differentiated features include: (1) FE_4 nests FE_1 on and FE_2 / FE_3 have the co-existence relationship with it, and FE_1 itself is categorised as *Deleted*; (2) FE_1 nests on S so that S is *Updated*; and (3) the other features having non-interacting relationships are *Unchanged*. In Fig. 8.6(f), the differentiation process of the faces in the *Updated* FE_1 is shown.

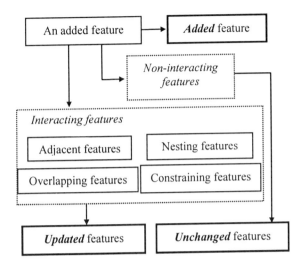

(a) *Differentiation process for an added feature*

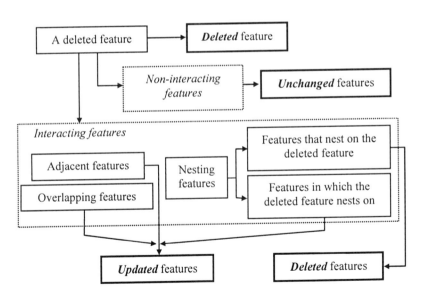

(b) *Differentiation process for a deleted feature*

Fig. 8.4 Feature manipulation operations in the environment.

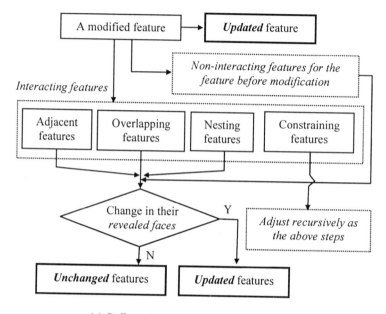

(c) *Differentiation process for a modified feature*

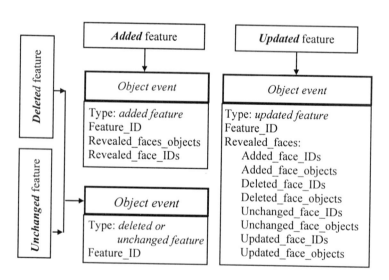

(d) *Object events to wrap differentiated features*

Fig. 8.4 Feature manipulation operations in the environment (cont'd).

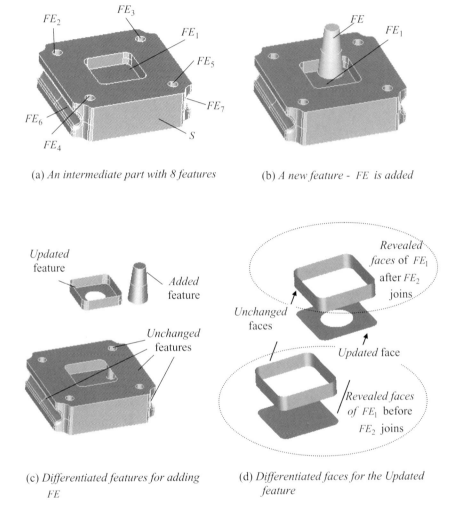

(a) *An intermediate part with 8 features*

(b) *A new feature - FE is added*

(c) *Differentiated features for adding FE*

(d) *Differentiated faces for the Updated feature*

Fig. 8.5 An example of the feature manipulation operations: a feature is added.

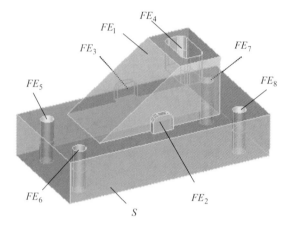

(a) *An intermediate part with 9 features*

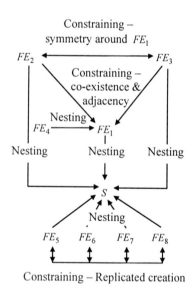

(b) *F2F_Graph of the features in the part*

Fig. 8.6 Examples of the feature manipulation operations: two features are deleted respectively.

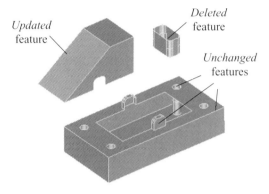

(c) *Differentiated features for deleting* FE_4

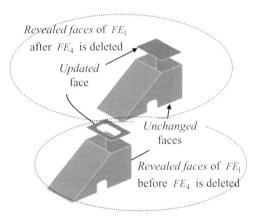

(d) *Differentiated faces for the Updated feature*

Fig. 8.6 Examples of the feature manipulation operations: two features are deleted respectively (cont'd).

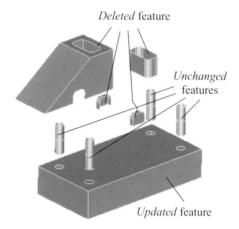

(e) *Differentiated features for deleting* FE_1

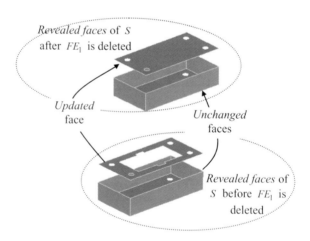

(f) *Differentiated faces for the Updated feature*

Fig. 8.6 Examples of the feature manipulation operations: two features
are deleted respectively (cont'd).

In Fig. 8.7, two features in a part are edited and the differentiation processes are shown. Features in the part and the *F2F_Graph* of these features are illustrated in Fig. 8.7(a) and (b). A feature, FE_{13}, in the part is edited and its interacting features are highlighted in Fig. 8.7(c). The classified features are shown in Fig. 8.7(d), in which the *Updated* features include FE_{13}, FE_1, FE_2 and S, and the other features that have non-interacting relationships with FE_{13} are *Unchanged*. With the changes to the position parameters of FE_9 in Fig. 8.7(e), its constrained features, FE_{10}, FE_{11} and FE_{12}, and the adjacent features, FE_5 / FE_6, and FE_9 itself are categorised as *Updated*, while the other features that have not interacted with the updated FE_9 are *Unchanged* in Fig. 8.7(f).

8.3 Distributed and Collaboration Mechanisms

8.3.1 *Communication mechanism for distributed design*

In this research, the establishment of the distributed design environment is based on the Java RMI. According to the RMI mechanism, through declaring remote interfaces, methods inherited from the interfaces and implemented can be used for remote calling and transmitting information. Some remote interfaces defined in the environment are shown in Fig. 8.8.

A three-layer inheritance mechanism for defining events is designed to take advantage of the object-oriented concept and provide a structural and extensible way to wrap the various types of information that are communicated in the distributed environment. According to Java specifications, to communicate in a network, the defined events should be "Serialisable". In the first layer, a supper class for events, which inherits the "Serialisable" class of the Java language, is defined and its sub-classes in the other two layers are automatically "Serialisable". On the other hand, the upper class provides a unified event variable sent or received by the remote methods declared in the remote interfaces of the distributed environment so as to simplify the system structure.

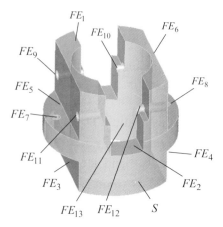

(a) *An intermediate part with 14 features*

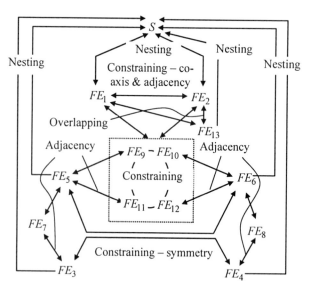

(b) *F2F_Graph of the features in the part*

Fig. 8.7 Examples of the feature manipulation operations: two features are
        deleted respectively.

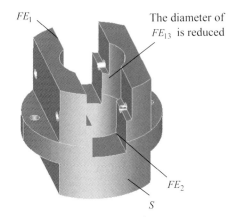

(c) *An edition operations on a feature - FE_{13}*

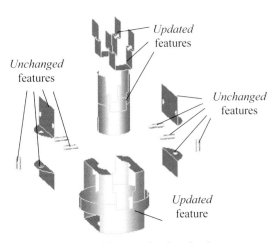

(d) *Differentiated features for the edited FE_{13}*

Fig 8.7 Examples of the feature manipulation operations: two features
are edited respectively (cont'd).

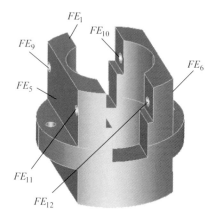

(e) *An edition operations on another feature -* FE_9

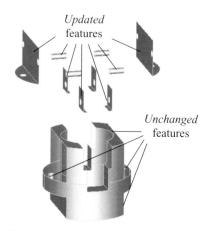

(f) *Differentiated features for the edited* FE_9

Fig 8.7 Examples of the feature manipulation operations: two features are edited respectively (cont'd).

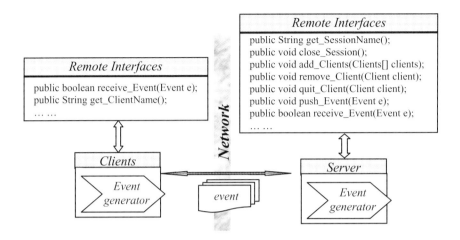

Fig. 8.8 Remote interfaces in the distributed environment.

The event classes in the second layer, which extends the supper event class, are mainly classified into the following three types:

(1) Parameter event for a design feature. This event, which is generated in a client, is used to wrap the parameters of a feature or a set of selected *revealed faces* for local operations on an existing feature. This event is dispatched to the server for creating a feature represented as a B-Rep object.

(2) Object event for a design feature. This event wraps the *revealed faces* in a feature and one or more events (since one or more features might be involved due the editing of a feature) from the server are sent back to the clients for visualisation and manipulation.

(3) Object event for a design part. The features of a design part are wrapped in this event to be dispatched from a client to the CAPP module for analysis.

The procedures of invoking remote methods are unidirectional in a basic RMI mechanism, i.e., a client must look up a server and call its remote methods. In an Intranet environment, in order to enable clients to update design information only when the server has a new event to communicate, instead of routinely pinging the server for information and creating a network backlog, a "call-back" mechanism can be employed to achieve a high-performance and robust server activity. The working process based on the call-back mechanism is described as follows and depicted in Fig. 8.9:

(1) A list is created in a working session maintained by the session manager to store the references of design clients that have joined the session.

(2) With an input of parameters for a feature, a parameter event is generated in a client. Through invoking one of the server methods - *push_Event(Event e)*, such an event is received and handled by the server. After an object event has been created and is ready for broadcasting from the server, each client recorded in the reference list is activated to receive the event by invoking one of the clients' remote methods - *receive_Event()*.

Firewalls, which are applied between an enterprise's Intranet and the Internet as the sheltering confines of the internal information, block all network traffic beyond the Intranet with the exception of those intended for certain well-known ports, such as the HTTP 80 port. Due the mechanism of the dynamic socket connections, the RMI traffic is typically blocked by most of firewalls. To address this problem, a "HTTP tunnelling" mechanism has been introduced by Sun Inc. to encapsulate RMI calls within an HTPP POST request to go cross the 80 port. In this case, the call-back mechanism is de-activated and replaced by the basic RMI mechanism, which is pinging the server from the clients for information updating, which is usually up to 10 times slower. Hence, depending on the condition of whether a client is inside or outside of the firewall of the server, the call-back- or HTTP tunnelling-based working process are available to choose from.

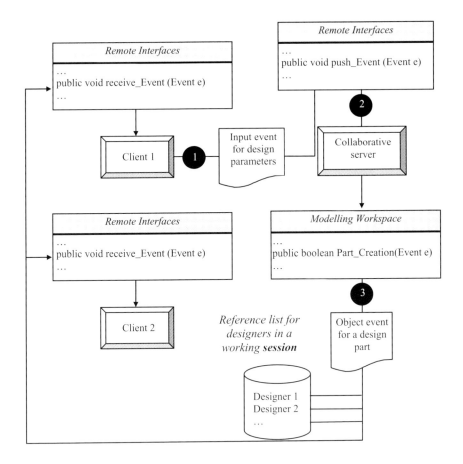

Fig. 8.9 A "call-back" process for clients (designers) and the collaborative server to communicate.

8.3.2 *Mechanism for collaborative design*

The process of designing a part collaboratively in the environment is depicted in Fig. 8.10. On the server side, a working session can be dynamically created and accessed by clients to provide a workspace to carry out collaborative design activities, in which clients can play different roles and take on different responsibilities. Designers participating in the same session can share the same design model.

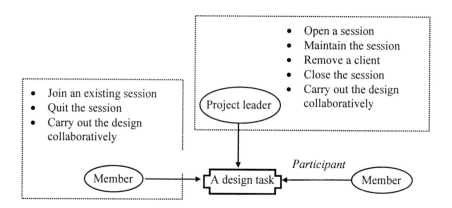

Fig. 8.10 Process of carrying out a design task in the environment.

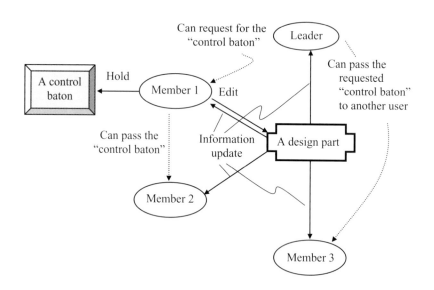

Fig. 8.11 The control process of designing a part through a "control baton".

Within a session, a "control baton" mechanism is employed to control and schedule the collaborative design activity. Each session has a control baton, i.e., at any one time, only the user who holds the control baton is the active designer and can edit a part; while the other users in the same session can only receive the updated information and are observers. The user who is carrying out an editing function can become an observer by transferring his control baton to another user. A project leader supervises the whole design process. This project leader is authorised to schedule the process to avoid unreasonable monopoly of the control baton and deadlocks due to network problems. The design process based on a control baton mechanism is shown in Fig. 8.11.

8.3.3 *A case study*

A practical part is used to illustrate the co-design processes in the distributed design environment. Fig. 8.12(a) shows the viewer on the client side and the design part in it. During the modelling process by a designer, the relevant intermediate information is packaged as events and shared with other designers automatically in a design session through the event-driven and call-back mechanisms. In the viewer on the client side, each designer has the freedom to adjust the viewing properties of the part, such as the colour, viewing position and background, for his/her visualisation convenience and preference.

Fig. 8.12(b) shows a discussion pad and a session manager in the environment. In the session manager, designers can log on/off a session and the control baton can be exchanged. During the working process of a designer, a discussion pad can be invoked by any other designers in the design session to make some comments or discussions based on a captured picture of the design part. Designers can chat through text or label the picture for sharing ideas.

Fig. 8.12(c) shows the features and the *PC_Tree* of the part, in which FE_4 - FE_8 consist of several individual features respectively, and these individual features have constraints with each other to constitute compound features.

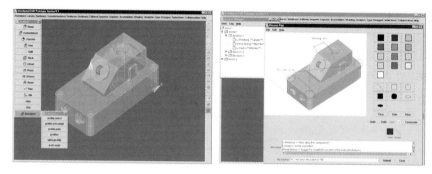

(a) A design part shown in a client   (b) *The session manager and discussion pad*

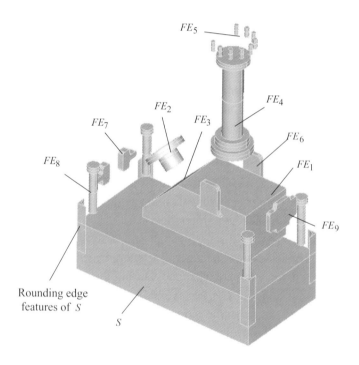

(c) *Features and the PC_Tree of the part*

Fig. 8.12 A case study for designing a part in the distributed environment.

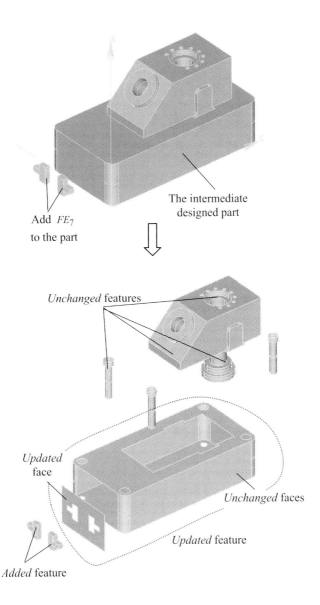

Fig. 8.13 The differentiation process for the adding of FE_7 in the intermediate part.

A designer holds the "control baton" and models the part. For the part, the intermediate process for adding FE_7 is shown in Fig. 8.13.

The size of the B-Rep of the part is approximately 964k, while the size of the face-based representation of the part is reduced to 595k. The size of the *Added* and *Updated* faces in the features, which are the major information in the object events for features to transmit from the server to the clients, is about 255k. Three parameters are defined to represent the data reductions from three perspectives:

$$\begin{cases} A1 = \text{face - based representation/B - Rep;} \\ A2 = \text{Added and Updated faces/face - based representation;} \\ A3 = \text{Added and Updated face/B - Rep.} \end{cases}$$

Among them, A3 reflects the actual reduction percentage of the reduced communication traffic for a part in B-Rep from the server side to the client sides. For the intermediate part at this stage, A1, A2 and A3 are approximately 61.7%, 42.9% and 26.5% respectively. A1, A2 and A3 for the intermediate processes are shown in Fig. 8.14, and their mean values are approximately 70%, 40% and 30% respectively. More experiments have been performed to show that by utilising the proposed feature manipulation operations and the feature representation mechanism, the reduction percentage of the reduced communication traffic for the design parts (A3) in the environment is roughly about 20% - 40% of the original data in the B-Reps.

After the first designer finishes the modelling process, another designer acquires the control baton and makes a modification on the part. This second designer thickened FE_6 to enhance its strength. FE_6 has two constrained features and the adjustment of a feature will affect the other. FE_6 is adjacent to S and FE_1. During the modification of FE_6, the *revealed faces* of S are updated while the *revealed faces* in FE_1 are kept unchanged. Hence, the *Updated* features in this stage include FE_6 and S, and the other features are *Unchanged*. A1, A2 and A3 are 58.5%, 13.8% and 6.3% respectively. Some results are illustrated in Fig. 8.15.

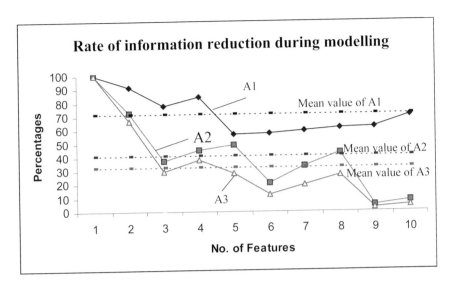

Fig. 8.14 The reduction rates of the represented and transmitted information.

8.4 Summary

In this chapter, a client/server environment has been developed based on feature and Java technologies to enable a dispersed team to accomplish a feature-based design task collaboratively. In the environment, a "manipulation client + modelling server" infrastructure has been developed to facilitate consistency of primary modelling information for multiple users and adaptability of the system to the web applications in the future. A distributed feature mechanism has been proposed to filter the varied information of a working part during a co-design activity to avoid unnecessary re-transferring of the complete large-size CAD files each time when any interactive operation is imposed on the model by a client, so as to enhance the effectiveness of the information communication for co-design activities.

In the distributed environment, a design task and engaged clients are organised and connected through working sessions generated and maintained dynamically by a collaborative server. The information from an individual client during a design process is updated and broadcast to

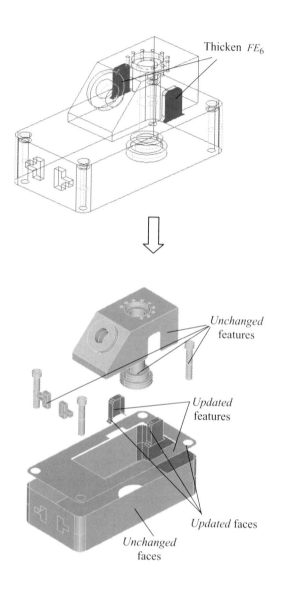

Fig. 8.15 The differentiation process for the edited FE_6 in the part.

other clients in the same session through an event-driven mechanism based on the RMI. The environment is open to downstream manufacturing analysis modules to achieve distributed concurrent engineering principle. A case study is used to illustrate a collaborative design activity and show that the proposed environment can facilitate a practical teamwork for distributed design with high-performance communications and organisation.

Bibliography

Products, Prototypes and Standards wit Websites (last access: 5/Oct/2004):

ABLE (Agent Building and Learning Environment), www.research.ibm.com/able.

ACISHOOPS 3D part viewer ™, www.hoops3d.com.

Actify SpinFire™, www.actify.com.

Adaptive Media Envision3D™, www.adaptivemedia.com.

Aglets, www.trl.ibm.com/aglets.

Alibre Design™, www.alibre.com.

Autodesk Streamline™, www.autodesk.com.

BS Contact™, www.bitmanagement.de.

CATIA™, www.catia.com.

Centric Software Pivotal Studio™, www.centrisoftware.com.

Cimmetry Systems Autovue™, www.cimmetry.com.

CollabCAD™, www.collabcad.com.

ConceptStation™, www.realitywave.com.

ConceptWorks™, www.realitywave.com.

Contact™, www.blaxxun.com.

Cortona™, www.parallelgraphics.com.

Cosmo Player™, www.cosmosoftware.com.

Enovia™, www.ibm.com/software/appliations/plm/enovia.

E-vis™, www.evis.com.

FIPER Project, www.fiperproject.com/fiperindex.htm.

Flux™, www.mediamachines.com.

HOOPS stream control™/stream plug-in™, www.hoops3d.com.

Hoops Streaming Toolkit™, hoops3d.com/products/hoops/stream.htm.

Inventor™, www.autodesk.com.

IX SPeeD™, www.impactxoft.com/products/suiteV5.asp.

JADE (JAva DEvelopment), Sharon.cselt.it/projects/jade.

JATLite (Java Agent Template Lite), java.stanford.edu.

Onespace™, www.onespace.net.

OpenHSF, www.openhsf.org.

313

ParaHOOPS 3D part viewer™, www.hoops3d.com.
ProjectLink™, www.ptc.com/products/windchill/projectlink.htm.
SmarTeam™, www.smarteam.com.
SolidEdge™, www.solidedge.com.
SolidWorks eDrawing™, www.solidworks.com.
SolidWorks™, www.solidworks.com.
TeamCenter™, www.ugs.com/products/teamcenter.
Unigraphics™, www.ugs.com.
VRWeb, www2.iicm.edu/vrweb.
W3D (Web 3D), www.macromedia.com.
X3D (eXtensible 3D), www.x3d.com.
XGL/ZGL, www.xglspec.org.
Xj3D, www.xj3d.org.
ZEUS, www.labs.bt.com/prjects/agents.htm.

Aarts, E. and Korst, J. (1989). *Simulated Annealing and Boltzman Machine*, John Wiley & Sons.

Allamaraju, S., et al. (2001). *Professional Java Server Programming – J2EE 1.3 Edition*, Wrox Press.

Bailey, M.W. and VerDuin, W.H. (2000). FIPER – an intelligent system for the optimal design of highly engineered products, *NIST Performance Metrics for Intelligent Systems Workshop*, Gaithersburg, MD, USA.

Balamuralikrishna, R., Athinarayanan, R. and Song, X.S. (2000). The Relevance of concurrent engineering in industrial technology programs, *Journal of Industrial Technology*, 16(3), pp. 1-5.

Baumgart, B.G. (1974). *Geometric Modelling for Computer Vision*, PhD Thesis, Stanford University, USA.

Begole, J., Struble, C.A. and Shaffer, C.A. (1997). Leveraging Java applets: toward collaboration transparency in Java, *IEEE Internet Computing*, 1-2, pp. 57-64.

Beiter, K.A. and Ishii, K. (2003). Integrating producibility and product performance tools within a Web-service environment, *Proceedings of ASME 2003 Design Engineering Technical Conferences*, Chicago, IL, USA, DETC03/CIE-48281.

Bezdek, E.J., Thompson, D.C., Wood, K.L. and Crawford, R.H. (1999). Volumetric feature recognition for direct engineering, *Direct Engineering: Toward Intelligent Manufacturing (Edited by Kamrani, A.K. and Sferro, P.R.)*, Kluwer Academic Publishers, Boston, pp. 16-69.

Bezier, P. (1966). Definition numerique des courbes et surfaces I, *Automatisme*, 11, pp. 625-632 (in French).

Bianconi, F. and Conti, P. (2003). Collaborative product modelling in heterogeneous environments: an approach based on XML schema, *Proceedings of the 10th ISPE*

International Conference on Concurrent Engineering: Research and Applications, July 26-30, Portugal, pp. 303-310.

Brownsboro, W.F. and Noort, A. (2004). Multiple-view feature modelling for integral product development, *Computer-Aided Design*, 36(10), pp. 929-946.

Burkett, W.C. (2001). Product data mark-up language: a new paradigm for product data exchange and integration, *Computer-Aided Design*, 33(7), pp. 489-500.

Carpenter, G.A. and Grossberg, S. (1987). ART2: self-organization of stable category recognition codes for analog input patterns, *Applied Optics*, 26(23), pp. 4919-4930.

Case, M.P. and Lu, S.C.Y. (1995). Discourse model for collaborative design, *Computer-Aided Design*, 28(5), pp. 333-345.

Chamberlain, M.A., Joneja, A. and Chang, T.C. (1993). Protrusion-features handling in design and manufacturing planning, *Computer-Aided Design*, 25(1), pp. 19-28.

Chang, T.C. (1990). *Expert Process Planning for Manufacturing*, Addison-Wesley, New York, USA.

Chao, K.M., Norman, P., Anane, R. and James, A. (2002). An agent-based approach to engineering design, *Computers in Industry*, 48(1), pp. 17-27.

Chen, C.L.P. and Leclair, S.R. (1994). Integration of design and manufacturing: solving setup generation and feature sequencing using an unsupervised-learning approach, *Computer-Aided Design*, 26(1), pp. 59-75.

Chen, J., Zhang, Y.F. and Nee, A.Y.C. (1998). Set-up planning using Hopfield net and simulated annealing, *International Journal of Production Research*, 36(4), pp. 981-1000.

Chen, L., Song, Z.J. and Feng, L. (2004). Internet-based real-time collaborative assembly modelling via an e-assembly system: status and promise, *Computer-Aided Design*, 36(9), pp. 835-847.

Chen, Y.M. and Liang, M.W. (2000). Design and implementation of a collaborative engineering information system for allied concurrent engineering, *International Journal of Computer Integrated Manufacturing*, 13(1), pp. 11-30.

Cheng, K., Pan, P.Y. and Harrison, D.K. (2001). Web-based design and manufacturing support systems: implementation perspectives, *International Journal of Computer Integrated Manufacturing*, 14(1), pp. 14-27.

Choi, B.K., Barash, M.M. and Anderson, D.C. (1984). Automatic recognition of machined surfaces from a 3D solid model, *Computer-Aided Design*, 16(2), pp. 81-86.

Choi, H.J., Panchal, J., Allen, J.K., Rosen, D. and Mistree, F. (2003). Towards a standardized engineering framework for distributed, collaborative product realization, *Proceedings of ASME 2003 Design Engineering Technical Conferences*, Chicago, IL, USA, DETC03/CIE-48279.

Chu, C.C.P. and Gadh, R. (1996). Feature-based approach for set-up minimization of process design from product design, *Computer-Aided Design*, 28(5), pp. 321-332.

Chuang, S.H. (1991). *Feature Recognition from Solid Models Using Conceptual Shape Graphs*, PhD Thesis, Arizona State University, Tempe, USA.

Coons, S. (1964). *Surfaces for Computer-Aided Design*, MIT Technical Report, Available as AD 663 504 from the National Technical Information Service, Springfield, USA.

Corney, J. and Clark, D.E.R. (1991). Method for finding holes and pockets that connect multiple faces in a 2.5D objects, *Computer-Aided Design*, 23(10), pp. 658-668.

Cox, M. (1972). The numerical evaluation of B-Splines, *Journal of the Institute of Mathematics and Its Application*, 10, pp. 134-149.

Dagli, C.H. (1994). *Artificial Neural Networks for Intelligent Manufacturing*, Chapman & Hall.

Dagli, C.H. and Sittisathanchai, A. (1993). Genetic neuro-scheduler for job shop scheduling, *Computers & Industrial Engineering*, 25(1/4), pp. 267-270.

de Boor, C. (1972). On calculating with B-Splines, *Journal of Approximation Theory*, 6(1), pp. 50-62.

de Kraker, K.J., Dohmen, M. and Bronsvoort, W.F. (1995). Maintaining multiple views in feature modelling, *Proceedings of Solid Modelling'95 – Third Symposium on Solid Modelling and Applications*, Salt Lake City, Utah, USA, pp. 123-130.

de Kraker, K.J., Dohmen, M. and Bronsvoort, W.F. (1997). Maintaining multiple views in feature modelling, *Proceedings of Symposium on Solid Modelling and Applications 1997*, New York, USA, pp. 123-130.

De Martino, T., Falcidieno, B. and Habinger, S. (1993). Integration of design-by-features and feature recognition approaches through a unified model, *Modelling in Computer Graphics Methods and Applications*, Springer-Verlag, pp. 423-437.

De Martino, T., Falcidieno, B. and Habinger, S. (1998). Design and engineering process integration through a multiple view intermediate modeller in a distributed object-oriented system environment, *Computer-Aided Design*, 30(6), pp. 437-452.

DeFloriani, L. (1989). Feature Extraction from Boundary Models of 3D Objects, *IEEE Pattern Analysis and Machine Intelligence*, 11(8), pp. 785-798.

Dong, J. and Vijayan, S. (1997a). Features extraction with the consideration of manufacturing processes, *International Journal of Production Research*, 35(8), pp. 2135-2155.

Dong, J. and Vijayan, S. (1997b). Manufacturing feature determination and extraction: part II - A heuristic approach, *Computer-Aided Design*, 29(7), pp. 475-484.

Dong, J. and Vijayan, S. (1997c). Manufacturing feature determination and extraction: part I - Optimal volume segmentation, *Computer-Aided Design*, 29(6), pp. 427-440.

Faheem, W., Castano, J.F., Hayes, C.C. and Gaines, D.M. (1998). What is a manufacturing interaction? *Proceedings of 1998 ASME Design Engineering Technical Conferences*, Atlanta, Georgia, USA, pp. 1-6.

Falcidieno, B. and Giannini, F. (1989). Automatic recognition and representation of shape-based features in a geometric modelling system, *Computer Vision, Graphics, and Image Processing*, 48(1), pp. 93-123.

Fausett, L.V. (1994). *Fundamentals of Neural Networks*, Prentice Hall, New York.

Field, M.C. and Anderson, D.C. (1993). Fast feature extraction for machining applications, *Computer-Aided Design*, 26(11), pp. 803-813.

Fussell, S.R., Kraut, R.E., Lerch, F.J., Scherlis, W.L., McNally, M.W. and Cadiz, J.J. (1998). Coordination, overload and team performance: effects of team communication strategies, *Proceedings of the ACM Conference on Computer-Supported Cooperative Work*, November 14-18, Seattle, Washington, USA, pp. 275-284.

Gadh, R. and Prinz, F.B. (1992). Recognition of geometric forms using the differential depth filter, *Computer-Aided Design*, 24(11), pp. 583-598.

Gadh, R. and Sonthi, R. (1998). Geometric shape abstractions for internet-based virtual prototyping, *Computer-Aided Design*, 30(6), pp. 473-486.

Gaines, D.M. and Hayes, C.C. (1999). Custom-Cut: a customisable feature recogniser, *Computer-Aided Design*, 31(2), pp. 85-100.

Gao, S. and Shah, J.J. (1998). Automatic recognition of interacting machining features based on minimal condition sub-graph, *Computer-Aided Design*, 30(9), pp. 727-739.

Gayanker, P. and Henderson, M.R. (1990). Graph-based extraction of protrusions and depressions from boundary representations, *Computer-Aided Design*, 22(7), pp. 442-450.

Gera, C.D., Regli, W.C., Braude, I., Shapirstein, Y. and Foster, C.V. (2002). A collaborative 3D environment for authoring design semantics, *IEEE Computer Graphics and Applications*, 22(3), pp. 43-55.

Gerhard, J.F., Rosen, D., Allen, J.K. and Mistree, F. (2001). A distributed product realization environment for design and manufacturing, *ASME Journal of Computing and Information Science in Engineering*, 1(3), pp. 235-244.

Glover, F. (1997). *Tabu Search*, Kluwer Academic Publishers.

Gordon, W.J. and Riesenfeld, R.F. (1974). B-Spline curves and surfaces, *Computer-Aided Geometric Design (edited by Barnhill, R.E. and Riesenfeld, R.F.)*, Academic Press, New York, USA, pp. 95-126.

Grefenstette, J. (1987). Incorporating problem specific knowledge into genetic algorithms, *Genetic Algorithms and Simulated Annealing (Edited by Davis, L.)*, Kaufman, CA, USA, pp. 42 - 60.

Gu, Z. (1997). *Development of a Generic Process Planning System*, PhD Thesis, National University of Singapore, Singapore.

Gu, Z., Zhang, Y.F. and Nee, A.Y.C. (1995). Generic form feature recognition and operation selection using connectionist modelling, *Journal of Intelligent Manufacturing*, 6, pp. 263-273.

Gu, Z., Zhang, Y.F. and Nee, A.Y.C. (1997). Identification of important features for machining operations sequence generation, *International Journal of Production Research*, 35(8), pp. 2285-2307.

Gutwin, C. and Greenberg, S. (1999). The effects of workspace awareness support on the usability of real-time distributed groupware, *ACM Transactions on Computer-Human Interaction*, 6(3), pp. 243-281.

Han, J. and Requicha, A.A.G. (1997). Integration of feature based design and feature recognition, *Computer-Aided Design*, 29(5), pp. 393-403.

Han, J. and Requicha, A.A.G. (1998(a)). Feature recognition from CAD models, *IEEE Computer Graphics and Applications*, 18(3/4), pp. 80-94.

Han, J. and Requicha, A.A.G. (1998(b)). Modeller independent feature recognition in a distributed environment, *Computer-Aided Design*, 30(6), pp. 453-463.

Han, J., Regli, W.C. and Brooks, S. (1998). Hint-based reasoning for feature recognition: status report. Technical note 4 of special panel session for feature recognition at the 1997 ASME CIE conference, *Computer-Aided Design*, 30(13), pp. 1003-1007.

Haykin, S. (1999). *Neural Networks: a Comprehensive Foundation*, Prentice Hall.

Henderson, M.R. (1984). *Extraction of Feature Information from Three Dimensional CAD Data*, PhD Thesis, Purdue University, West Lafayette, USA.

Hoffmann, C.M. and Joan-Arinyo, R. (2000), Distributed maintenance of multiple product views, *Computer-Aided Design*, 32(7), pp. 421-431.

Huang, G.Q. (2002). Web-based support for collaborative product design review, *Computers in Industry*, 48(1), pp. 71-88.

Huang, G.Q., Lee, S.W. and Mak, K.L. (1999). Web-based product and process data modelling in concurrent "design for X", *Robotics and Computer-Integrated Manufacturing*, 15(1), pp. 53-63.

Hwang, J.L. and Henderson, M.R. (1992). Applying the perceptron to three-dimensional feature recognition, *Journal of Design and Manufacturing*, 2(4), pp. 187-198.

Ibrhim, R.N. and McCormack, A.D. (2002). Process planning using adjacency-based feature extraction, *International Journal of Advanced Manufacturing Technology*, 20(11), pp. 817-823.

Irani, S.A., Koo, H.Y. and Raman, S. (1995). Feature-based operation sequence generation in CAPP, *International Journal of Production Research*, 33(1), pp. 17-39.

Ishibuchi, H., Yamamoto, N., Murata, T. and Tanaka, H. (1994). Genetic algorithms and neighborhood search algorithms for fuzzy flowshop scheduling problems, *Fuzzy Sets and Systems*, 67(1), pp. 81-100.

Jacquel, D. and Salmon, J. (2000). Design for manufacturability: a feature-based agent-driven approach, *Journal of Engineering Manufacture, Proceedings of the Institution of Mechanical Engineers Part B*, 214(10), pp. 865-880.

Jha, K. and Gurumoorthy, B. (2000). Multiple feature interpretation across domains, *Computers in Industry*, 42(1), pp. 13-32.

Ji, Q. and Marefat, M.M. (1995). Bayesian approach for extracting and identifying features, *Computer-Aided Design*, 27(6), pp. 435-454.

Joshi, S. and Chang, T.C. (1988). Graph-based heuristics for recognition of machined from a 3D solid model, *Computer-Aided Design*, 20(2), pp. 58-66.

Kailash, S.B., Zhang, Y.F. and Fuh, J.Y.H. (2001). A volume decomposition approach to machining feature extraction of casting and forging components, *Computer-Aided Design*, 33(8), pp. 605-617.

Kan, H.Y., Duffy, V.G. and Su, C.J. (2001). An Internet virtual reality collaborative environment for effective product design, *Computers in Industry*, 45(2), pp. 197-213.

Kim, Y.S. (1990). *Convex Decomposition and Solid Geometric Modelling*, PhD Thesis, Stanford University, USA.

Kim, Y.S. and Wang, E. (2002). Recognition of machining features for cast then machined parts, *Computer-Aided Design*, 34(1), pp. 71-87.

Kim, Y.S. and Wilde, D.J. (1992). A convergent convex decomposition of polyhedral objects, *Transactions of ASME, Journal of Mechanical Design*, 114, pp. 468-476.

Kirkpatrick, S., Gelatt, Jr C.D. and Vecchi M.P. (1983). Optimisation by simulated annealing, *Science*, 220, pp. 671-680.

Kong, S.H., Noh, S.D., Han, Y.G., Kim, G. and Lee, K.I. (2002). Internet-based collaborative system: press-die design process for automobile manufacturer, *International Journal of Advanced Manufacturing Technology*, 20(9), pp.701-708.

Kotak, D., Wu, S.H., Fleetwood, M. and Tamoto, H. (2003). Agent-based holonic design and operations environment for distributed manufacturing, *Computers in Industry*, 52(2), pp. 95-108.

Kruth, J.P. and Detand, J. (1992). A CAPP system for non-linear process plans, *Annals of the CIRP*, 41(1), pp. 489-492.

Kyprianou, I.K. (1980). *Shape Classification in Computer-Aided Design*, PhD Thesis, University of Cambridge, UK.

Laakko, T. and Mantyla, M. (1993). Feature modelling by incremental recognition, *Computer-Aided Design*, 25(8), pp. 479-492.

Lang, S.Y.T., Dickinson, J. and Buchal, R.O. (2002). Cognitive factors in distributed design, *Computers in Industry*, 48(1), pp. 89-98.

Lankalapalli, K., Chatterjee, S. and Chang, T.C. (1997). Feature recognition using ART2: a self-organizing neural network, *Journal of Intelligent Manufacturing*, 8(3), pp. 203-214.

Lee, D.H., Kiritsis, D. and Xirouchakis, P. (2001). Search heuristics for operation sequencing in process planning, *International Journal of Production Research*, 39(16), pp. 3771-3788.

Lee, J.Y. and Kim, K. (1998). A feature-based approach to extracting machining features, *Computer-Aided Design*, 30(13), pp. 1019-1035.

Lee, J.Y. and Kim, K. (1999). Generating alternative interpretations of machining features, *International Journal of Advanced Manufacturing Technology*, 15(1), pp. 38-48.

Lee, J.Y., Kim, H., Han, S.B. and Park, S.B. (1999). Network-centric feature-based modelling, *Proceedings of the Seventh Pacific Graphics Conference on Computer Graphics and Applications*, October 5-7, South Korea, pp. 280-289.

Lee, Y.C. and Fu, K.S. (1987). Machine Understanding of CSG: extraction and unification of manufacturing features, *IEEE Computer Graphics and Application*, 7(1), pp. 20-32.

Li, W.D., Ong, S.K., Fuh, J.Y.H., Wong, Y.S., Lu, Y.Q. and Nee, A.Y.C. (2004(a)). Feature-based design in a collaborative and distributed environment, *Computer-Aided Design*, 36(9), pp. 775-797.

Li, W.D., Fuh, J.Y.H. and Wong, Y.S. (2004(b)). An Internet-enabled integrated system for co-design and concurrent engineering, *Computers in Industry*, 55(1), pp. 87-103.

Li, W.D., Ong, S.K. and Nee, A.Y.C. (2000). Recognition of overlapping machining features based on hybrid artificial intelligence techniques, *Proceedings of the Institution of Mechanical Engineers Part B, Journal of Engineering Manufacture*, 214, pp. 739 - 744.

Li, W.D., Ong, S.K. and Nee, A.Y.C. (2002). Recognizing manufacturing features from a design-by-feature model, *Computer-Aided Design*, 34(11), pp. 849-868.

Li, W.D., Ong, S.K. and Nee, A.Y.C. (2003). A hybrid method for recognizing manufacturing features, *International Journal of Production Research*, 2003, 41(9), pp. 1887-1908.

Li, W.D., Lu, W.F., Fuh, J.Y.H. and Wong, Y.S. (2005(a)). Collaborative computer-aided design – research and development status, *Computer-Aided Design* , in press.

Li, W.D., Ong, S.K. and Nee, A.Y.C. (2005(b)). A Web-based process planning optimization system for distributed design, *Computer-Aided Design* , in press.

Lin, A.C., Lin, S.Y., Diganta, D. and Lu, W.F. (1998). An integrated approach to determining the sequence of machining operations for prismatic parts with interacting features, *Journal of Materials Processing Technology*, 73, pp. 234-250.

Liu, X.D. (2000). CFACA: component framework for feature-based design and process planning, *Computer-Aided Design*, 32(7), pp. 397-408.

Ma, G.H., Zhang, Y.F. and Nee, A.Y.C. (2000). A simulated annealing-based optimization algorithm for process planning, *International Journal of Production Research*, 38(12), pp. 2671-2687.

Mantyla, M. (1988). *Introduction to Solid Modelling*, Computer Science Press, New York, USA.

Marefat, M. and Kashyap, P.L. (1990). Geometric reasoning for recognition of three-dimensional object features, *IEEE Transactions on Pattern Analysis and Machine Intelligence*, 12(10), pp. 949-965.

Mathias, K.E., Whitley, L.D., Stork, C. and Kusuma, T. (1994). Staged hybrid genetic search for seismic data imaging, *Proceedings of the First IEEE Conference on Evolutionary Computation*, Orlando, USA, pp. 356-361.

Mervyn, F., Senthil kumar, A., Bok, S.H. and Nee, A.Y.C. (2003). Development of an Internet-enabled interactive fixture design system, *Computer-Aided Design*, 35(10), pp. 945-957.

Mintzberg, H. (1983). *Structures in Fives: Designing Efficient Organizations*, Prentice-Hall, Englewood Cliffs, New Jersey, USA.

Mori, T. and Cutkosky, M.R. (1998). Agent-based collaborative design of parts in assembly, *Proceedings of 1998 ASME Design Engineering Technical Conferences*, Atlanta, Georgia, USA, DETC98/CIE-5697.

Nam, T.J. and Wright, D.K. (1998). CollIDE: a shared 3D workspace for CAD, *Proceedings of the 4th EATA International Conference on Network Entities*, October 15-16, Leeds, UK, pp. 389-400.

Nezis, K. and Vosniakos, G. (1997). Recognizing 2.5D shape features using a neural network and heuristics, *Computer-Aided Design*, 29(7), pp. 523-539.

Ong, S.K. and Nee, A.Y.C. (1998). Manufacturing evaluation and generation of re-design suggestions for machined parts, *International Journal for Manufacturing Science & Production*, 1(2), pp. 87-105.

Onwubolu, G.C. (1999). Manufacturing features recognition using backpropagation neural networks, *Journal of Intelligent Manufacturing*, 10(3/4), pp. 289-299.

Owodunni, O. and Hinduja, S. (2002a). Evaluation of existing and new feature recognition algorithms Part 1: theory and implementation, *Proceedings of the Institution of Mechanical Engineers Part B, Journal of Engineering Manufacturing*, 216(6), pp. 839-851.

Owodunni, O. and Hinduja, S. (2002b). Evaluation of existing and new feature recognition algorithms Part 2: experimental results, *Proceedings of the Institution of Mechanical Engineers Part B, Journal of Engineering Manufacturing*, 216(6), pp. 853-866.

Ozturk, N. and Ozturk, F. (2001). Neural network based non-standard feature recognition to integrate CAD and CAM, *Computers in Industry*, 45(2), pp. 123-135.

Pallot, M. and Sandoval, V. (1998). *Concurrent Enterprising: Toward the Concurrent Enterprise in the Era of the Internet and Electronic Commerce*, Kluwer Academic Publishers.

Perng, D.B. and Chang, C.F. (1997). Resolving feature interactions in 3D part editing, *Computer-Aided Design*, 29(10), pp. 687-699.

Pham, D.T. and Karaboga, D. (2000). Intelligent Optimisation Techniques: Genetic Algorithms, Tabu Search, Simulated Annealing and Neural Networks, Springer, London, UK.

Prabhakar, S. and Henderson, M.R. (1992). Automatic form-feature recognition using neural-network-based techniques on boundary representation of solid models, *Computer-Aided Design*, 24(7), pp. 381-393.

Qiang, L., Zhang, Y.F. and Nee, A.Y.C. (2001). A distributed and collaborative concurrent product design system through the WWW/internet, *International Journal of Advanced Manufacturing Technology*, 17(5), pp. 315-322.

Qiao, L., Wang, X.Y. and Wang, S.C. (2000). A GA-based approach to machining operation sequencing for prismatic parts, *International Journal of Production Research*, 38(14), pp. 3283-3303.

Ranky, P.G. (1994). Current/Simultaneous Engineering: Methods, Tools & Case Studies, CIMware Limited, Guildford, Surrey, UK.

Rao, S.S., Nahm, A., Shi, Z.Z., Deng, X.D. and Syamil, A. (1999). Artificial intelligence and expert systems applications in new product development – a survey. *Journal of Intelligent Manufacturing*, 10, pp. 231-244.

Reddy, S.V.B., Shunmugam, M.S. and Narendran, T.T. (1999). Operation sequencing in CAPP using genetic algorithms, *International Journal of Production Research*, 37(5), pp. 1063-1074.

Regli, W.C. (1995). *Geometric Algorithm for Recognition of Features from Solid Models*, PhD Thesis, University of Maryland, College Park, USA.

Regli, W.C., Gupta, S.K. and Nau, D.S. (1997). Towards multiprocessor feature recognition, *Computer-Aided Design*, 29(1), pp. 37-51.

Renders, J. and Bersini, H. (1994). Hybridizing genetic algorithms with hill-climbing methods for global optimization: two possible ways, *Proceedings of the First IEEE Conference on Evolutionary Computation*, Orlando, USA, pp. 312-317.

Rogers, D. (1991). G/SPLINES: A hybrid of Friedman's multivariate adaptive regression splines (MARS) algorithm with Holland's genetic algorithm, *Proceedings of the Fourth International Conference on Genetic Algorithms*, San Diego, CA, pp. 384-391.

Roy, U. and Kodkani, S.S. (1999). Product modelling within the framework of the world wide web, *IIE Transactions*, 31(7), pp. 667-677.

Saad, M. and Maher, M.L. (1996). Shared understanding in computer-supported collaborative design, *Computer-Aided Design*, 28(3), pp. 183-192.

Sakurai, H. (1995). Volume decomposition and feature recognition: Part 1 - polyhedral objects, *Computer-Aided Design*, 27(11), pp. 833-843.

Sakurai, H. (1996). Volume decomposition and feature recognition: Part 2 - curved objects, *Computer-Aided Design*, 28(6/7), pp. 519-532.

Sakurai, H. and Gossard, D.C. (1990). Recognizing shape features in solid models, *IEEE Computer Graphics and Applications*, 10(9), pp. 22-32.

Schoenberg, I. (1946). Contributions to the problem of approximation of equidistant data by analytic functions, *Quarterly Applied Mathematics*, 4, pp. 45- 99.

Sentil kumar, A., Salim, F.K. and Nee, A.Y.C. (1996). Automatic recognition of design and machining features from prismatic parts, *International Journal of Advanced Manufacturing Technology*, 11, pp. 136-145.

Shah, J.J., Shen, Y. and Shirur, A. (1994). Determination of machining volumes from extensible sets of design features, *Advances in Feature Based Manufacturing (Edited by Shah, J.J., Mantyla, M. and Nau, D.S.)*, Elsevier/North-Holland, Amsterdam, pp. 129-158.

Shen, W.M., Maturana, F. and Norrie, D.H. (2000). MetaMorph II: an agent-based architecture for distributed intelligent design and manufacturing, *Journal of Intelligent Manufacturing*, 11, pp. 237-251.

Shyamsundar, N. and Gadh, R. (2002). Collaborative virtual prototyping of product assemblies over the Internet, *Computer-Aided Design*, 34(10), pp. 755-768.

Sommerville, M., Clark, D. and Corney, J. (1995). Viewer-centred feature recognition, *Proceedings of Solid Modelling '95 – Third Symposium on Solid Modelling and Applications*, Salt Lake City, Utah, USA, pp. 125-129.

Suh, H. and Ahluwalia, R.S. (1995). Feature modification in incremental feature generation, *Computer-Aided Design*, 27(8), pp. 627-635.

Sun, J., Zhang, Y.F. and Nee, A.Y.C. (2001). A distributed multi-agent environment for product design and manufacturing planning, *International Journal Production Research*, 39(4), pp. 625-645.

Sung, H.A., et al. (2001). CyberCut: an Internet-based CAD/CAD system, Transactions of the ASME, Journal of Computing and Information Science in Engineering, 1, pp. 52-59.

Sutherland, I.E. (1974). *Sketchpad: A Man-Machine Graphical Communication System*, PhD Thesis, MIT, Cambridge, USA.

Tay, F.E.H. and Roy, A. (2003). CyberCAD: a collaborative approach in 3D-CAD technology in a multimedia-supported environment, *Computers in Industry*, 52(2), pp. 127-145.

Teti, R. and Kumara, S.R.T. (1997). Intelligent computing methods for manufacturing systems, *Annals of CIRP*, 46(2), pp. 629-650.

Trika, S.N. and Kashyap, R.L. (1994). Geometric reasoning for extraction of manufacturing features in iso-oriented polyhedrons, *IEEE Transactions on Pattern Analysis and Machine Intelligence*, 16(11), pp. 1087-1100.

Tseng, Y.J. and Joshi, S. (1994). Recognizing multiple interpretations in 2.5D machining of pockets, *International Journal of Production Research*, 32(5), pp. 1063-1086.

Tseng, Y.J. and Lin, C.C. (1998). Analysis on multiple sets of feature-based tool paths for prismatic machining parts, *International Journal of Production Research*, 36(12), pp. 3491-3509.

Tseng, Y.J. and Liu, C.C. (2001). Concurrent analysis of machining sequences and fixturing set-ups for minimizing set-up changes for machining mill-turn parts, *International Journal of Production Research*, 39(18), pp. 4197-4214.

Tuttle, R., Little, G., Corney, J. and Clark, D.E.R. (1998). Feature recognition for NC part programming, *Computers in Industry*, 35(3), pp. 275-289.

van den Berg, E., Bidarra, R. and Bronsvoort, W.F. (2000). Web-based interaction on feature models, *Proceedings of the Seventh IFIP WG 5.2 Workshop on Geometric Modelling: Fundamentals and Applications*, October 2-4, Parma, Italy, pp. 113-123.

Vancza, J. and Markus, A. (1991). Genetic algorithms in process planning, *Computers in Industry*, 17(3), pp. 181-194.

Vandenbrande, J.H. and Requicha, A.A.G. (1993). Spatial reasoning for the automatic recognition of machinable features in solid models, *IEEE Transactions on Pattern Analysis and Machine Intelligence*, 15(12), pp. 1269-1285.

Venuvinod, P.K. and Wong, S.Y. (1995). A graph-based expert system approach to geometric feature recognition, *Journal of Intelligent Manufacturing*, 6(3), pp. 155-162.

Versprille, K. (1975). Computer-Aided Design Applications of the Rational B-Spline Approximation Form, PhD Thesis, Syracuse University, New York, USA.

Voelker, H. and Requicha, A.A.G. (1977). Geometric modelling of mechanical parts and processes, *Computer*, 10(12), pp. 48-57.

Wang, E. and Kim, Y.S. (1998). Form feature recognition using convex decomposition: results presented at the 1997 ASME CIE Feature Panel Session, *Computer-Aided Design*, 30(13), pp. 983-988.

Wang, L., Shen, W., Xie, H., Neelamkavil, J. and Pardasani, A. (2002). Collaborative conceptual design: a state-of-the-art survey, *Computer-Aided Design*, 34(13), pp. 981-996.

Wang, M.T. and Chang, T.C. (1990). Feature recognition for automated process planning, *Proceedings of the ASME Manufacturing International Conference*, New York, pp. 49-55.

Wong, S.T.C. (1997). Coping with conflict in cooperative knowledge-based systems, *IEEE Transactions on Systems, Man and Cybernetics – Part A: Systems and Humans*, 27(1), pp. 57-72.

Wong, T.N. and Lam, S.M. (2000). Recognition of machining features – a hybrid approach, *International Journal of Production Research*, 38(17), pp. 4301-4316.

Wong, T.N. and Siu, S.L. (1995). A knowledge-based approach to automated machining process selection and sequencing, *International Journal of Production Research*, 33(12), pp. 3465-3484.

Woo, T.C. (1977). Computer-aided recognition of volumetric design, in *Advances in Computer-Aided Manufacturing (Edited by McPherem, M.)*, North-Holland, pp. 121-136.

Woo, T.C. (1982). Feature extraction by volume decomposition, *Proceedings of the Conference on CAD/CAM technology in Mechanical Engineering*, Cambridge, MA, USA, pp. 76-94.

Woo, Y. and Sakurai, H. (2002). Recognition of maximal features by volume decomposition, *Computer-Aided Design*, 34(3), pp. 195-207.

Wu, D. and Sarma, R. (2001). Dynamic segmentation and incremental editing of boundary representations in a collaborative design environment, *Proceedings of the 6th ACM Symposium on Solid Modelling and Applications*, June 4-8, Michigan, USA, pp. 289-300.

Wu, D. and Sarma, R. (2004). The incremental editing of faceted models in an integrated design environment, *Computer-Aided Design*, 36(9), pp. 821-833.

Wu, H.C. and Chang, T.C. (1998). Automated set-up selection in feature-based process planning, *International Journal of Production Research*, 36(3), pp. 695-712.

Wu, M.C. and Liu, C.R. (1996). Analysis on machined feature recognition techniques based on B-Rep, *Computer-Aided Design*, 28(8), pp. 603-616.

Xie, Y.L. and Salvendy, G. (2003). Agent-based features for CAD browsers in foster engineering collaboration over the Internet, *International Journal of Production Research*, 41(16), pp. 3809-3829.

Xu, X. and Hinduja, S. (1998). Recognition of rough machining features in 2.5D components, *Computer-Aided Design*, 30(7), pp. 503-516.

Yen, J., Liao, J.C. and Lee, B. (1998). A hybrid approach to modelling metabolic systems using genetic algorithm and simplex method, *IEEE Transactions on Systems, Man, and Cybernetics*, Part B, 28(2), pp. 173-191.

Yip-Hoi, D. and Dutta, D. (1996). A genetic algorithm application for sequencing operations in process planning for parallel machining, *IIE Transactions*, 28, pp. 55-68.

Yui, W. and Egbelu, P.J. (2000). Process alternative generation from production geometric design data, *IIE Transactions*, 32, pp. 71-82.

Zhang, C., Chan, K.W. and Chen, Y.H. (1998). A hybrid method for recognizing feature interactions, *Computer Integrated Manufacturing Systems*, 9(2), pp. 120-128.

Zhang, F., Zhang, Y.F. and Nee, A.Y.C. (1997). Using genetic algorithms in process planning for job shop machining, *IEEE Transactions on Evolutional Computation*, 1(4), pp. 278-289.

Zhang, H.C. and Huang, S.H. (1994). A fuzzy approach to process plan selection, *International Journal of Production Research*, 32(6), pp. 1265-1279.

Zhang, S.S., Shen, W.M. and Ghenniwa, H. (2004). A review of Internet-based product information sharing and visualisation, *Computers in Industry*, 54(1), pp. 1-15.

Zhao, F.L., Tso, S.K. and Wu, P.S.Y. (2000). A cooperative agent modelling approach for process planning, *Computers in Industry*, 41(1), pp. 83-97.

Zhou, S.Q., Chin, K.S., Xie, Y.B. and Yarlagadda, P. (2003). Internet-based distributive knowledge integrated system for product design, *Computers in Industry*, 50(2), pp. 195-205.

Zulkifli, A.H. and Meeran, S. (1999). Feature patterns in recognizing non-interacting and interacting primitive, circular and slanting features using a neural networks, *International Journal of Production Research*, 37(13), pp. 3063-3100.

Index

327